CEC5R08

DIAGNOSTIC TOOLS TO EVALUATE IMPACTS OF TRACE ORGANIC COMPOUNDS

FINAL REPORT

by:

Jerry Diamond, Ph.D.
Tetra Tech, Inc.

Kent Thornton, Ph.D.
FTN Associates, Inc.

Kelly Munkittrick, Ph.D.
Karen Kidd, Ph.D.
University of New Brunswick

Steve Bartell, Ph.D.
E2 Consulting Engineers

Katherine Kapo, Ph.D.
Montani Run, LLC

2011

WERF

The Water Environment Research Foundation, a not-for-profit organization, funds and manages water quality research for its subscribers through a diverse public-private partnership between municipal utilities, corporations, academia, industry, and the federal government. WERF subscribers include municipal and regional water and wastewater utilities, industrial corporations, environmental engineering firms, and others that share a commitment to cost-effective water quality solutions. WERF is dedicated to advancing science and technology addressing water quality issues as they impact water resources, the atmosphere, the lands, and quality of life.

For more information, contact:
Water Environment Research Foundation
635 Slaters Lane, Suite G-110
Alexandria, VA 22314-1177
Tel: (571) 384-2100
Fax: (703) 299-0742
www.werf.org
werf@werf.org

This report was co-published by the following organization.

IWA Publishing
Alliance House, 12 Caxton Street
London SW1H 0QS, United Kingdom
Tel: +44 (0) 20 7654 5500
Fax: +44 (0) 20 7654 5555
www.iwapublishing.com
publications@iwap.co.uk

Library of Congress Catalog Card Number: 2010940626

IWAP ISBN: 978-1-84339-547-8/1-84339-547-9

Tetra Tech, Inc.

ACKNOWLEDGMENTS

The authors wish to thank all of the individuals who helped develop this document. In addition, the authors also wish to thank the project subcommittee and program director for their valuable input, which helped to make this a useful product for WERF subscribers and the water resource community as a whole.

Report Preparation

Principal Investigator:
Jerry Diamond, Ph.D.
Center for Ecological Sciences, Tetra Tech, Inc.

Research Team:
Kent Thornton, Ph.D.
FTN Associates, Inc.

Steve Bartell, Ph.D.
E2 Consulting Engineers, Inc.

Kelly Munkittrick, Ph.D.
Karen Kidd, Ph.D.
University New Brunswick

Katherine Kapo, Ph.D.
Montani Run LLC

Jennifer Flippin
Herb Brass, Ph.D.
Henry Latimer
Jaime Gilliam
Jeroen Gerritsen, Ph.D.
Center for Ecological Sciences, Tetra Tech, Inc.

Technical Review Committee:

Scott Dyer, Ph.D.
The Proctor and Gamble Company

Drew McAvoy, Ph.D.
University of Cincinnati

Erich Emery
Ohio River Sanitation Commission

Steve Bay, Ph.D.
Southern California Coastal Water Research Project

Paul DeLeo, Ph.D.
American Cleaning Institute

Deb Lester
King County Department of Natural Resources

Margaret Nellor, P.E.
Nellor Environmental Associates, Inc.

Jim Pletl
Hampton Roads Sanitation District

Elizabeth Toot-Levy
Northeast Ohio Regional Sewer District

Mary Buzby, Ph.D.
Merck & Co.

Mike Focazio
U.S. Geological Survey

Water Environmental Research Foundation Staff:

Director of Research: Daniel M. Woltering, Ph.D.
Program Director: Lola Olabode, MPH

ABSTRACT AND BENEFITS

Abstract:

With the recent advent of improved analytical and biomarker detection capabilities, a variety of organic chemicals have been found in trace amounts (Trace Organic Chemicals, TOrCs) in surface waters and fish tissue. TOrCs include pharmaceuticals, personal care products, surfactants, pesticides, flame retardants, and other organic chemicals, some with unknown modes of action or effects. Identifying or predicting ecological effects of TOrCs in typical aquatic multi-stressor situations is challenging, requiring a variety of epidemiological tools that together, can diagnose effects at multiple scales of ecological organization. The goal of this research is to provide information on TOrCs to help the water quality community make scientifically defensible and cost effective decisions that are appropriately protective of aquatic populations and communities. Five objectives were addressed in this research: 1) develop and apply a procedure to prioritize which TOrCs are of most concern; 2) develop and test a conceptual site screening framework to determine if sites are or could be affected by TOrCs; 3) evaluate and test diagnostic approaches to identify potential risks due to TOrCs using various case studies; 4) develop a relational database and user interface with which the water resource community can enter, store, and search TOrC exposure and occurrence data in the U.S.; and 5) foster partnerships and transfer knowledge gained in this research to the water quality community. TOrC fate, effects, and occurrence data were compiled in a database for over 500 organic chemicals based on over 100 published studies representing more than 50 organizations and 700 sites. Alternative risk-based prioritization processes and draft lists of high priority TOrCs were developed. A preliminary site screening and diagnostic framework was developed and evaluated using seven different case study sites. EPA's causal analysis (stressor identification) procedures, Canada's Environmental Effects Monitoring (EEM) procedure, the ecosystem model CASM (Comprehensive Aquatic System Model), and several other specialized diagnostic tools were used and evaluated. A relational database based on Tetra Tech's EDAS2 was developed using the Microsoft platform. The modified version of EDAS2, built on the EPA WQX data model, provides web-based data queries using a combination of tabular data for downloads and a visual map interface that allows the user to view, query, and select sites from the map having chemical or biological data. This final report summarizes all approaches used and results obtained in this research; discusses critical data gaps and other important uncertainties, and provides testable hypotheses and recommendations for Phase 2 testing and analyses.

Benefits:

- Provides potentially useful prioritization approaches for identifying TOrCs that should be monitored or considered at a given location.
- Collates useful information regarding fate and effects of TOrCs monitored by numerous water quality agencies.
- Provides an approach to calculate estrogenicity equivalents of various TOrCs for prioritization and diagnostic assessments.

- Identifies potentially useful screening indicators to help prioritize sites with respect to ecological risks due to TOrCs.
- Demonstrates potentially useful diagnostic approaches and types of population and community-level analyses as well as other analyses to identify causes of ecological impairments in aquatic systems due to TOrCs.
- Identifies critical biological data needs to evaluate whether TOrCs are causing or could cause biological impairment.
- Provides a compilation of diagnostic approaches and types of analyses used to identify causes of ecological impairments in aquatic systems.
- Provides a conceptual approach for diagnosing aquatic ecological effects due to TOrCs.

Keywords: Contaminants of emerging concern, trace organic chemical, endocrine disruptors, aquatic ecology, stressor identification.

TABLE OF CONTENTS

LIST OF TABLES

LIST OF FIGURES

LIST OF ACRONYMS

BOD	Biological oxygen demand
CADDIS	Causal Analysis Diagnostic Decision Information System
CAFO	Concentrated Animal Feeding Operation
CSO	Combined Sewer Overflow
DO	Dissolved oxygen
EAL	Estrogenic activity level
ECOSAR	Ecological Structure Activity Relationships
EDC	Endocrine disruptor chemical
EE2	17-α ethynylestradiol
EEM	Environmental Effects Monitoring
EPT	Ephemeroptera, Plecoptera, Trichoptera
ERA	Ecological Risk Assessment
EROD	Ethoxyresorufin-O-deethylase assay
FDA	Food and Drug Administration
GIS	Geographic Information System
HPG	Hypothalmus-Pituitary-Gonad axis
IBI	Index of biotic integrity
IOC	Investigation of cause
MOA	Mode of action
PAH	Polycyclic aromatic hydrocarbons
PBDE	Polybrominated diethylether
PCB	Polychlorinated biphenyl
PCS	Permit Compliance System
PPCP	Pharmaceuticals and Personal Care Products
QSAR	Quantitative Structural Activity Relationship
SSD	Species sensitivity distribution
TDS	Total dissolved solids
TSS	Total suspended solids
TOrC	Trace Organic Compound
USDA	U.S. Department of Agriculture
U.S. EPA	U.S. Environmental Protection Agency
USGS	U.S. Geological Survey
VTG	Vitellogenin
WERF	Water Environment Research Foundation
WWTP	Wastewater treatment plant

EXECUTIVE SUMMARY

With the advent of improved analytical and biomarker detection capabilities, a variety of organic compounds have been found in trace amounts (Trace Organic Compounds or TOrCs) in effluents, drinking water, surface waters, and fish tissue. TOrCs include pharmaceuticals, chemicals in personal care products, surfactants, current use pesticides, and flame retardants. Information regarding effects and mode of action (MOA) for some of these compounds is limited or non-existent. Identifying or predicting the ecological effects of TOrCs in typical surface waters is challenging, and requires a variety of tools to diagnose effects on aquatic populations or communities. The overall research challenge, of which this research is a part, is to understand and communicate potential exposure and effects from TOrCs. The goal of this and other WERF-sponsored TOrC research is to provide the essential frameworks, tools, and information to wastewater treatment managers to ensure that they are making the best and most appropriate treatment decisions to guarantee healthy aquatic ecosystems in receiving waters.

This document presents: 1) a compilation of TOrCs monitored by many different organizations in surface waters and effluents over the past 10 years, and three different approaches for prioritizing TOrCs of most concern for the purpose of developing priority TOrC monitoring lists; 2) a proposed framework for screening sites that may be at risk due to TOrCs, and 3) a generalized retrospective methodology that could be used to diagnose whether TOrCs cause or contribute to aquatic ecological effects at a site. Screening tools that could be applied either prospectively or retrospectively to assess ecological risks from TOrCs, are also presented (see Chapter 3.0).

The ultimate goal of this research is to develop a framework and tools for use by the water quality community to: (a) assess whether observed site-specific impacts on aquatic populations or communities are caused by TOrCs, and (b) predict whether measured exposure concentrations of TOrCs in effluent or surface water are likely to cause aquatic population and/or community effects. This provides a challenge as many higher levels of organization alone (population, community) are not diagnostic of impacts attributable only to specific TOrCs. Chapter 4.0 discusses the pros and cons of linking lower biological level diagnostic tools (biochemical, physiological, etc.) to higher levels of biological organization. Compared to biochemical endpoints, community endpoints have high ecological relevance, but a long time lag, low reversibility, and present challenges tying changes back to a specific cause. At the other end of the spectrum, biochemical responses have a short time lag, high reversibility, and good potential for linkage with causes, but have relatively low ecological relevance. Ultimately linking lower level changes to higher level changes may be important to definitively diagnosing TOrCs as the cause of community and population changes.

With shrinking budgets and a large universe of potential environmental concerns and issues, wastewater utilities and the water quality community at large need tools that help them prioritize risks and focus actions/expenditures on those issues that are most critical. The development of screening and diagnostic tools presented in this report could benefit end users by helping them prioritize which sites most require monitoring and assessment of TOrCs and whether TOrCs are a factor of concern at their site.

Figure ES-1 shows the general conceptualization of this research and the context of screening and diagnostic tools discussed in this report, which benefited from input received at an expert workshop held in 2010. Sites (e.g. a waterbody receiving treated effluent) should be screened first to determine whether they are low risk, high risk, or the risk is unknown for TOrC effects. If it can be demonstrated that the site is low risk for TOrC effects, further monitoring and assessment of TOrCs is probably not warranted. If the site could have a high risk (or unknown risk) due to TOrCs, monitoring may be warranted. In this case effluent or receiving waters could be screened using the developed priority TOrC lists to determine if high risk chemicals are present.

Task 1 of this research (Chapter 2.0) developed three types of approaches for prioritizing TOrCs that may be high risk and that might be considered for monitoring. Each approach has its advantages and limitations (Table ES-1). Using an occurrence database compiled from over 100 monitoring studies, the three prioritization approaches were applied to over 500 TOrCs that have been detected in water or effluent samples in the U.S. over the past 10 years. Some TOrC measurements reported and used in our occurrence database may not be accurate given the difficulties in obtaining reliable data for some types of TOrCs at low detection levels of interest. As the importance of emerging trace organic contaminants has been recognized relatively recently, methods have emerged with varying performance characteristics. While the quality of these methods has improved significantly, the performance of some are variable and actively being researched.

Acute and chronic toxicity information was compiled from EPA's ECOSAR and estrogenic activity of TOrCs was compiled using the Food and Drug Administration's database, which are compiled in a companion database for this project. In addition, fate information was compiled for these TOrCs using EPA's EpiSuite and PBT profiler, which is also available in the companion database. Acute and chronic toxicity endpoints derived from ECOSAR represent potential population-level effects. These endpoints are frequently used to develop water criteria for toxics for the protection of aquatic life populations and communities. Similar approaches could be taken to protect organisms from harm due to acute or chronic exposures to individual TOrCs. However, because individual TOrCs are not likely to occur in concentrations that are acutely toxic to aquatic life, alternative sublethal endpoints need to be considered. Endocrine activity endpoints, for example, are thought to be important to diagnosing population and community effects for some TOrCs.

Types of TOrCs identified as high priority differed among approaches: steroids, hormones, pharmaceuticals, and surfactants comprised most of the high priority TOrCs based on a risk approach, while pesticides, industrial chemicals, and PAHs comprised most of the high priority TOrCs based on a PBT approach (Figure ES-2). The hybrid approach predictably resulted in a mixture of high risk and high PBT TOrCs. Although it is not definitively known at this time, many of the high priority TOrCs using the risk-based approach could originate from treated wastewater sources, in addition to other sources, because these TOrCs are represented by many personal care products and pharmaceuticals that often associated with residential wastewater. TOrCs identified using the other two approaches are more likely to originate from non-wastewater treatment sources (e.g., nonpoint source runoff, CAFOs, agricultural practices). The risk-based approach resulted in the fewest number of high priority TOrCs (41) while the

PBT approach resulted in the greatest number of high priority TOrCs (108). A total of 23 TOrCs were identified as high priority using all three approaches and a total of 126 TOrCs were identified as high priority using any one of the approaches.

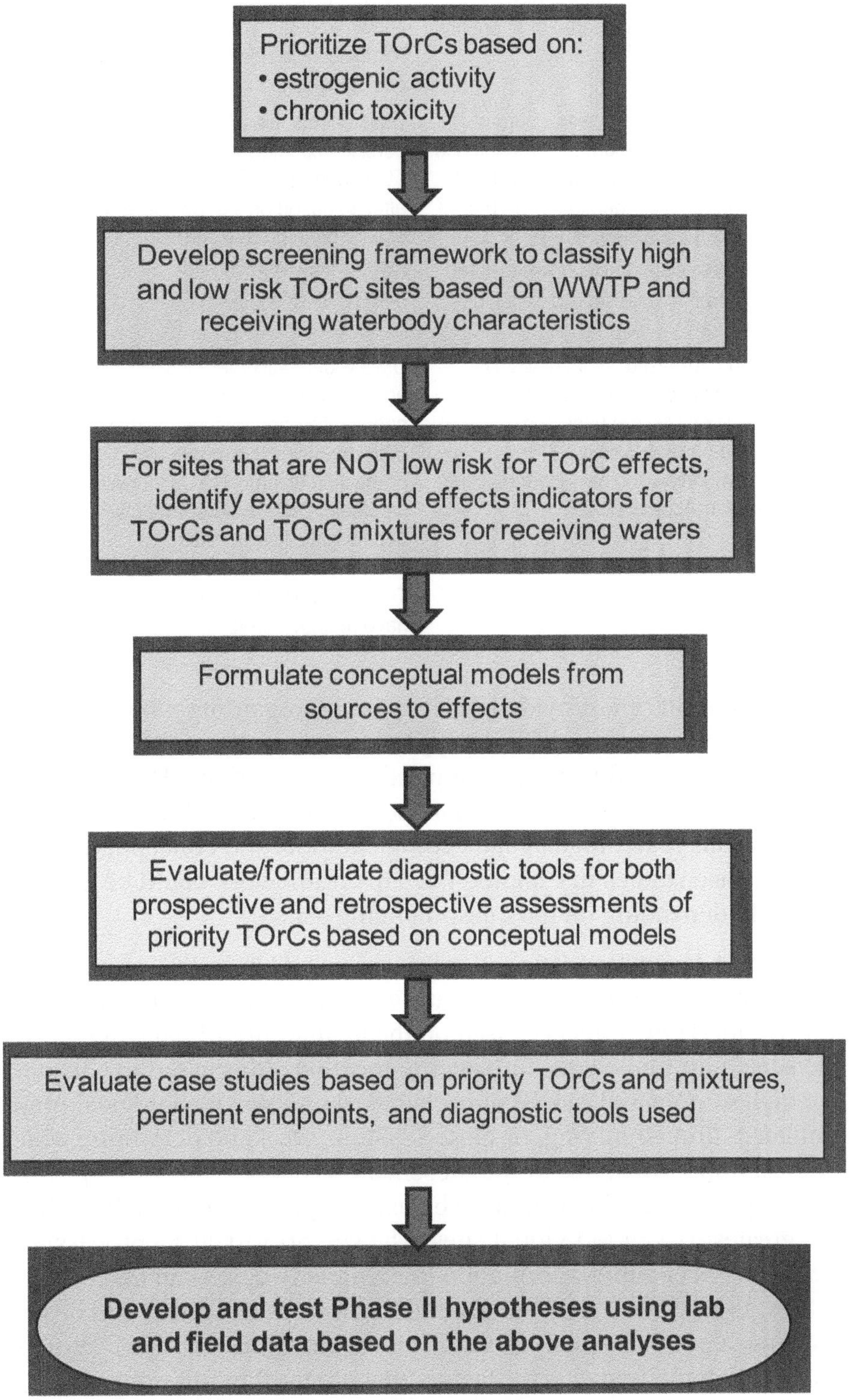

Figure ES-1. Flow Chart Showing the Steps Conducted in this Phase 1 Research.

Table ES-1. Advantages and Limitations of the Three Different Approaches Proposed for Prioritizing Trace Organic Chemicals (TOrCs).

Approach	Description	Advantages	Limitations
1	Risk-based: max concentration divided by most sensitive predicted effects threshold (toxicity or estrogenicity). Quotient ≥ 0.10 considered high priority.	Simple to use and communicate; based on actual measurements of TOrCs; considers TOrCs that are known to occur in surface waters; relatively high relevance to ecological risk.	Dependent on maximum concentrations observed and recorded in occurrence database; does not consider high production TOrCs not monitored.
2	Risk-based and fate-based: Sum of effect, bioaccumulation, and persistence scores; effect score based on quotient as in Approach #1; quotient ≥ 0.10 = score of 3 (highest priority); log K_{ow} ≥ 5.0 = score of 3; half-life in water ≥ 180 d = score of 3. Total score ≥ 7 is high priority TOrC.	Uses effects, bioaccumulation and persistence information for each TOrC-addresses fate as well as effects; based on TOrCs known to occur and that have been measured; relatively high relevance to ecological risk.	Dependent on TOrCs and their maximum concentrations in the occurrence database; does not consider HPV TOrCs not monitored.
3	Tox and fate-based: Sum of toxicity, bioaccumulation, and persistence scores; toxicity score based on predicted chronic toxicity; bioaccumulation and persistence scores same as in Approach #2.	Uses widely-accepted "PBT" approach-consistent chemical prioritization approaches used by other agencies; considers high production volume TOrCs as well as other TOrCs; not necessarily dependent on occurrence information; capable of identifying TOrCs that should be evaluated or have methods developed.	May include TOrCs that are unlikely to result in exposure to aquatic biota; based on toxicity effects, not endocrine effects; toxicity thresholds may have high uncertainty for some HPV TOrCs that don't have QSAR models.

Recommendations are provided in Chapter 2.0 regarding which TOrC prioritization approach might be best suited to different types of organizations (i.e., wastewater treatment plants, CAFOs, industrial effluents, etc) or monitoring questions. Having a manageable subset of high priority TOrCs, it is possible to evaluate potential effects of individual TOrCs as well as mixtures using various toxicological and ecological tools and conceptual models, such as those developed for the case studies in Chapter 4.0. This framework could be used both prospectively and retrospectively depending on the question being asked.

In addition to recommendations for prioritizing TOrCs, this report also develops a framework for prioritizing specific wastewater influenced sites to determine which locations may be at greatest risk for TOrC effects. Based on the availability of data and current knowledge, the preferred site screening framework for specific locations (discussed in Chapter 3.0), is to identify sites with the highest potential risk of impacts based on known ecological information (e.g., effluent dominated, limited mixing, or closed system, etc.) and determine which of these actually exhibit biological impairment due to the presence of TOrCs. The TOrC prioritization lists developed in Chapter 2.0 could eventually be used as guidance to predict which chemicals may be posing the greatest risk at individual sites. The site prioritization process has the advantage of relying on relatively accessible information that, in many cases can be obtained remotely through public databases. Figure ES-2 shows a general framework of factors and observations (biological, chemical, and physical) useful for screening sites for TOrC ecological risk, and Table ES-2 presents hypothesized risk factors that are used in the screening process. Further refinement of Table ES-2, and a scoring system for risk characteristics could be developed in Phase 2. Factors used in the site screening process are grouped as either: 1) wastewater influent characteristics, 2) wastewater treatment characteristics, 3) receiving system characteristics, and 4) site ecological and TOrC concentration data (measured or predicted). The criteria presented in this screening approach should be viewed as working hypotheses in this research and not as established risk drivers regarding wastewater sources of TOrCs.

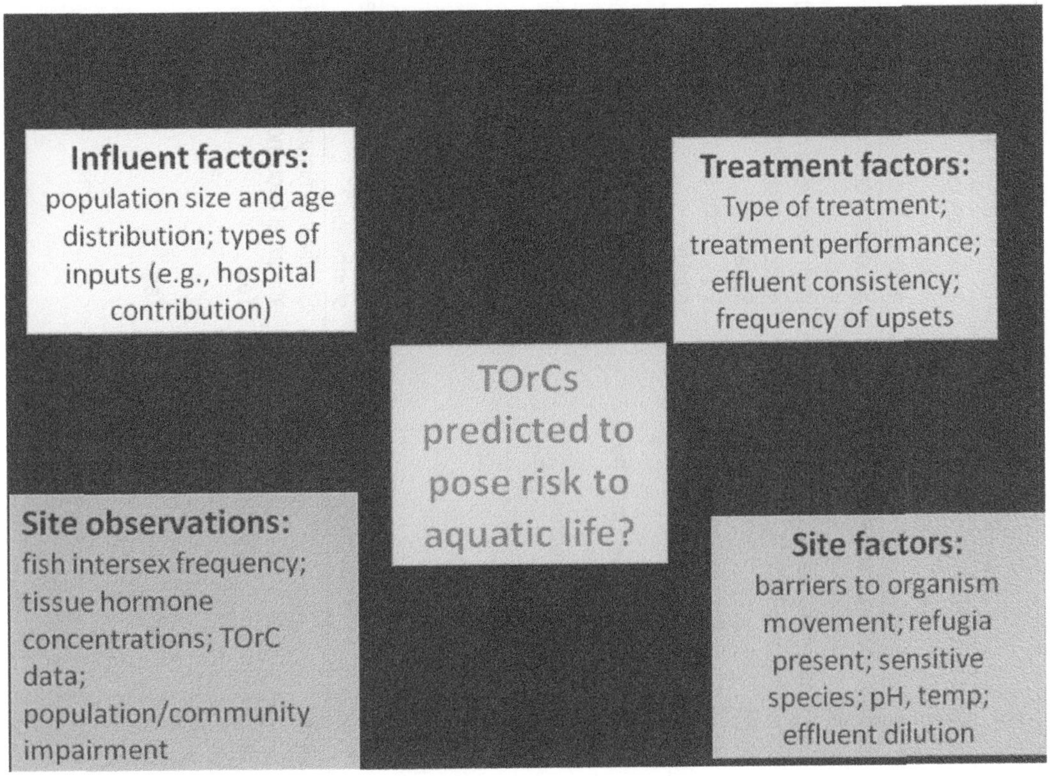

Figure ES-2. General Schematic Showing Hypothesized Factors Considered in Screening Wastewater Sites for Diagnostic Analyses Involving TOrCs.

At this point in the research, all factors in Table ES-2 should be considered when screening sites, although some factors may not be readily available for a given site. Ideally, missing information should be collected at specific sites to gain a better understanding of the ecological risks of TOrCs. If data are available indicating individual effects due to TOrCs, this should take precedence over any other factor in the screening process. If no site data exist, and predicted TOrC concentrations are below known toxicity or estrogenicity thresholds, all other screening factors should be evaluated if possible, prior to assigning the site "low priority" or "low risk potential" because the effect of TOrC mixtures is uncertain and often unknown. Once sites have been screened in terms of potential for risk due to TOrCs, more in-depth diagnostic tools can be used to determine whether TOrC exposure is sufficient to cause an unacceptable risk to aquatic biota. The use of site screening tools and other diagnostic tools are demonstrated using case studies described in Chapter 4.0.

Table ES-2. Draft Criteria for Screening Waste Water Treatment Plant (WWTP) Effluent-Influenced Sites for Potential Aquatic Ecological Effects Due to Trace Organic Chemicals.

	Factor	Risk Thresholds	
		Lower Risk	Higher Risk
WWTP Input Characteristics	Industry/hospital-related influents	None or very minor	>5% of influent
	Population served	< 10,000	> 1M
	Average age of population	< 30	> 50
WWTP Treatment Characteristics	Level of treatment	Tertiary, advanced secondary	Secondary or Primary
	Sludge retention time (SRT)	> 15d	< 5d
	Treatment effectiveness/uniformity	Low (near reference condition) nutrient concentrations; TSS	High (relative to reference conditions) nutrient levels; TSS
	WWTP upsets	None	Several
	Uniformity of influent flow and water quality	Relatively uniform most of the time	Very inconsistent in terms of flow and/or quality
Receiving Waterbody	Percent Effluent Instream (low flow)	< 1%	> 70%
	Waterbody size and/or habitat diversity	Large (e.g., > 5th order) waterbody; abundant refuges	Small waterbody; refuges few or non-existent
	Waterbody openness/barriers	No barriers to fish movement	Dams or other barriers preventing fish movement
	Threatened, endangered, or species of concern present	None	At least 1
	Presence of other potential sources of TOrCs	None	Agricultural (CAFOs, row crops, orchards) sources present upstream or nearby urban stormwater?
Site Observations	Existing EDC effects present	None	Significantly higher vtg induction in male fish; abnormal frequency of intersex in fish
	TOrC concentration > screening threshold level (from literature values)	No	Yes
	Impaired IBI/ICI community metric scores or low population counts	No	Yes

An example demonstrating application of the screening approach was conducted in a retrospective assessment of available data from wastewater-influenced sites in the Drift Plain ecoregion of Ohio (Chapter 4.0). This assessment used a GIS approach to integrate hypothesized site screening criteria, with modeled TOrC exposure concentrations and quantitative stressor-response associations generated from screening-level statistical analyses. The ultimate goal of this screening was to select sites showing impaired biology collocated with a WWTP, and few other stressors (e.g. agriculture, habitat, etc.). Cumulative risk scores (based on site information in comparison with the screening criteria) for wastewater treatment plant facilities and receiving waterbodies were determined based on summing risk criteria for each factor with available data, and facilities and receiving waterbodies having risk scores in the 25th and >75th percentile for Ohio were classified as "low risk" and "high risk", respectively. Sites of particular interest for TOrC risk had high risk criteria for both the associated wastewater treatment facility and receiving water characteristics, had predicted TOrC exposure (household chemicals and/or pharmaceuticals) identified as a potential stressor by quantitative analysis, and were either simple (TOrCs were the only predicted stressor) or complex (multiple stressors) risk scenarios (Table ES-3). This categorization of risk scenarios was used in several other case studies examined in this research. The use of these various lines of evidence in this screening assessment effectively utilized readily available data to screen high priority sites prior to devoting more extensive resources in assessment and diagnostics at the site level.

Table ES-3. TOrC Risk Scenario Categories Used in Screening Sites.

Site Risk classification	WWTP Facility	Receiving Waterbody
LEVEL I (Lowest)	Low risk	Low risk
LEVEL II	Low risk	High risk
LEVEL III	High risk	Low risk
LEVEL IV (Highest)	High risk	High risk

TOrC RISK (increasing from Level I to Level IV)

Two sites in Ohio that were screened as high risk for TOrCs, as well as five other sites from different geographic locations, were examined further to demonstrate the use of different diagnostic tools and to identify critical gaps in terms of diagnosing risks due to TOrCs (Table ES-4). Because most sites involve multiple potential stressors, causal analysis of ecological effects relies on strength of evidence using multiple lines of evidence. Criteria incorporated in EPA's CADDIS and Canada's EEM methodologies, are generally accepted lines of evidence for determining causality in retrospective assessments. However, as all of the case studies examined demonstrate, many of the potential lines of evidence often are not addressed. One of the critical challenges in diagnosing risks due to TOrCs is that at higher levels of organization (population, community), there may not be "diagnostic" impacts that are attributable only to specific TOrCs (or classes of TOrCs). A decline in fish growth rate, for example, could be due to a decline in availability of food resources, as much as to impaired physiology associated with chronic TOrC exposure. Therefore, the diagnostic approach must use multiple lines of evidence and look for patterns of responses that are not attributable to other stressors, as discussed in Chapter 4.0.

Table ES-4. Summary of Case Study Sites Examined.

Site (WWTP)	Location	Percent Effluent[1]	Potential Stressors in Addition to WWTP TOrCs	Biological Impairment?[2]	TOrC Data?	Scenario Type[3]	Risk Level[4]
SCB[5]	CA	<1	Legacy pesticides and PCBs; normal ocean temperature variation	Not Known	Water, Fish, Effluent, Sediment	A	I or III, depending on WWTP
Guelph	ON	40	Channelization; low-head dams; agricultural runoff	Not known	Water, Fish, Effluent	A	II
Kitchener	ON	10 - 15	Urban runoff; low-head dams; agricultural runoff; CAFOs[6]	Yes, fish	Water, Fish, Effluent	A	IV
Ravenna	OH	>90	Urban runoff; septic systems	Yes – fish, macroinvertebrates	Predicted	B	IV
Mansfield	OH	>80	Urban runoff; legacy industrial contaminants; industrial discharges; agricultural runoff	Yes – fish, macroinvertebrates; but upstream impaired too	Predicted	B	III
Taylor Run	PA	85	Urban runoff, stormwater	Yes – periphyton, macroinvertebrates	Fish Tissue	A	III
Boulder	CO	>70	Urban runoff; legacy mining pollutants; agricultural runoff	Not Known	Water, Fish, Effluent	A	Pre-upgrade= IV Post upgrade = II

[1] Under low flow or summer conditions

[2] Population or community level effects (e.g., changes in species abundance, biomass, age classes) observed downstream of the WWTP?

[3] Scenario categories based on Task 2 research in this project: A = TOrCs exceed conservative effect thresholds and instream effects have not been evaluated; B = Unknown whether TOrCs exceed thresholds and effects have been observed instream

[4] Risk levels based on screening tool developed in Task 2 of this project: Level I = low TOrC risk to aquatic biota in terms of wastewater treatment plant (WWTP) and instream factors; Level II = low risk based on WWTP factors but high risk based on instream factors; Level III = high risk based on WWTP factors but low risk based on instream factors; Level IV = high risk based on both WWTP and instream factors.

[5] Southern California Bight - Represents 4 WWTPs in and around Los Angeles

[6] Concentrated Animal Feeding Operation

While it is population/community effects that are of most interest from a risk assessment perspective (e.g., change in growth, reproductive potential, or abundance of a species), diagnosing risks due to TOrCs requires making links between effects at lower levels and those at higher levels. Morphological measures such as feminization of male fish are diagnostic of estrogenic TOrC exposure while the pattern of comparative weights of fish livers and gonads, combined with condition factor (i.e., Canada's EEM approach) may be diagnostic of different classes of stressors, including endocrine disruptors. These measures provided both negative evidence (i.e., TOrCs were unlikely to be a stressor at the Guelph site) and positive evidence (i.e., TOrCs were a likely contributor to impairment at the Kitchener site) in terms of TOrC diagnostics. Intersex and vitellogenin analyses were also instrumental in identifying estrogenic TOrCs as a stressor at the Boulder Creek site (Table ES-5). TOrC concentration data is important from a wastewater treatment plant perspective, to identify potential TOrC issues and identify potential treatment alternatives. However, TOrC data may not be informative in and of itself regarding actual TOrC exposure to biota. Organism physiological measures, such as vitellogenin concentration in male fish, hormone levels in fish, acetylcholinesterase activity in invertebrates, and other biomarkers may be more useful from the standpoint of establishing linkages between TOrC exposure and effects. A pivotal point in a TOrC diagnostic system is the individual organism (Figure ES-4). As demonstrated in several of our case studies, a number of physiological or histopathological measures may provide evidence that an individual organism has been exposed to certain TOrCs but there still needs to be a link made between individual exposure, or even exposure of multiple individuals, and measurable population-level effects. While certain types of TOrCs may have effects only on one assemblage (e.g., estrogenic TOrCs affect vertebrates such as fish), obtaining measures of effect for different assemblages is desirable to help diagnose whether stressors other than TOrCs may be responsible for observed effects. Rarely will TOrCs be the only potential stressors at a site, as shown in the case studies; the examination of multiple assemblages and ecological endpoints helped identify non-TOrC causes of impairment for the two Ohio sites and the West Chester, PA site. Both CADDIS and EEM provide a useful framework with which to evaluate multiple potential stressors.

In support of the research challenge for this Phase 1 effort, this project also developed a web-based relational database and user interface with which the water resource community could store and search TOrC data, as well as biological and aquatic life habitat information for sites in the U.S. Chapter 5.0 summarizes the purpose of the database, its structure and attributes, and ways in which the database and interface can be used. An initial database was created which includes TOrC data collected by several organizations over many sites across the U.S. The map tools provided with the user interface, along with several other search tools, allow a user to find TOrC data of interest, as well as metadata about specific TOrCs such as fate information (e.g., log Kow, half-life in water), obtained from EPA's EpiSuite, and toxicity and endocrine activity data, obtained from ECOSAR and FDA, respectively. This could be a useful tool for the water quality community to help track and analyze spatial and temporal trends in TOrC and associated ecological information at various locations.

Table ES-5. Summary of Available Information for Each of the Seven Case Studies, Relevant to Different Levels of Biological Organization and TOrC Diagnostic Analyses.

Site	Suborganism	Organism	Population	Community
SCB[1]	Vtg.[2] in male turbot; TOrCs and legacy CECs in livers; cortisol and thyroid hormones in blood	Ova-testis; intersex rate	Gender ratio; relative abundance	ND[3]
Guelph	TOrC analyses in water; Vtg., hormone, enzymes in fish exposed in situ	14-d in situ fish studies	Fish population studies	Macroinvertebrate and fish bioassessments
Kitchener	TOrC analyses in water; Vtg., hormone, enzymes in fish exposed in situ	Intersex rate;	Darter population studies	Macroinvertebrate and fish bioassessments
Ravenna	ND (predicted TOrC risk)	ND	ND	Macroinvertebrate and fish bioassessments
Mansfield	ND (predicted TOrC risk)	ND	ND	Macroinvertebrate and fish bioassessments
Taylor Run	TOrCs observed in fish tissue	ND	ND	Algal, macroinvertebrate, and fish bioassessments
Boulder Creek	Vtg. in male white suckers	Intersex in white suckers	Skewed sex ratio in white suckers	ND

[1] SCB – Southern California Bight
[2] Vtg. – vitellogenin analysis
[3] ND – No data

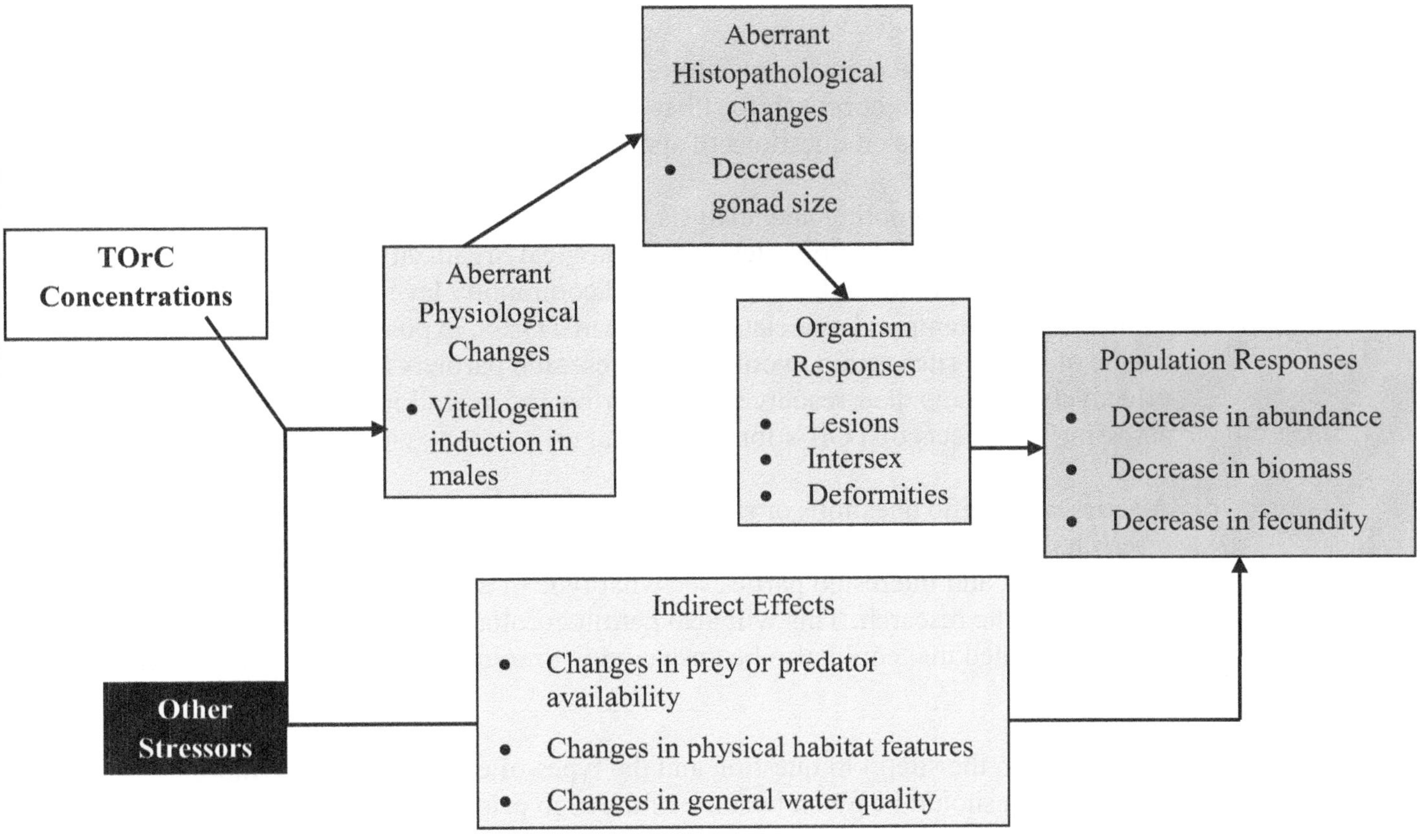

Figure ES-4. Relationships Between TOrC Exposure and Exposure to Other Stressors and Different Levels of Effect.

Recommendations regarding future monitoring and assessment activities that should be considered in Phase 2 of this project are discussed in full detail in Chapter 6.0 and include:

a. A continuum of research should be conducted in Phase 2. As evident from the Phase 1 research, establishing causation of TOrCs with impacts on aquatic organisms, and especially aquatic populations and communities, will require intensive, comprehensive, and time-consuming research. However, there are numerous intermediate questions and products on the way to causal answers that would be of benefit to WERF subscribers and partners. For example, being able to verify the Phase 1 screening approach for distinguishing potentially high risk from potentially low risk sites due to TOrCs could have significant benefits for WERF subscribers and partners. More information on individual risk parameters shown in Table ES-2 is necessary to determine the relative risk contribution of each characteristic. Once the relative importance of parameters is determined, a numerical risk scoring system could potentially be established for the parameters listed in Table ES-2. Low risk sites could be differentiated from high risk sites based on scores. In general, having better information on the associations of wastewater treatment and receiving stream factors with TOrC effects on individual organisms and aquatic populations, even if causation could not be unequivocally established, could significantly advance our understanding and management of TOrCs in the environment.

b. Nested designs of screening, associative, and intensive research sites would be a cost-effective and efficient approach for Phase 2 research and permit WERF partners and collaborators to focus on questions of specific concern to them at any of these sites.

c. Well-coordinated sampling and analysis, as well as co-located seasonal chemical and biological sampling at multiple levels of biological organization (i.e., physiological, histological, morphological, population, and community level metrics) is needed to evaluate spatial/temporal associations between TOrCs, exposure and effects. Having a suite of nested sites, again, would permit research partners and collaborators to cost-effectively leverage their resources and improve the overall information base for assessing the effects of TOrCs for everyone.

d. A series of specific questions or risk hypotheses should be formulated for the Phase 2 research that spans the range from screening to causation. This will permit partners, collaborators, and interested parties see what type of answers and products will be provided by the research. This will also permit corollary questions or risk hypotheses to be formulated that could also be answered with minimal or no additional resource requirements.

e. Depending on the site(s) in question and the types of effects evaluated, more consideration should be given to some of the high priority TOrCs identified in this research as opposed to legacy chemicals or those that necessarily have high persistence or bioaccumulation potential. This should also include mixtures of priority TOrCs. Many relatively labile chemicals may occur at a greater frequency than predicted based on their chemical structure because they are continuously introduced into the waterbody (e.g., via treated wastewater).

While uncertainties and data gaps still exist, utilities and other organizations managing water resources do not need to wait for Phase 2 to be completed in order to obtain useful information for assessing and managing TOrCs. The site screening tool developed in this research, while simple, is a useful first step to decide whether TOrCs are a potential cause for concern at a site. The prioritization approaches developed in this research are a useful starting point for deciding which TOrCs should be monitored in the receiving waterbody or effluent. Partitioning the ecological effects of trace level organic compounds on aquatic populations and communities will be a long term process. However, the Phase 1 research has provided tools for gathering information relating to population and community effects of TOrCs and laid a solid foundation for the Phase 2 research.

CHAPTER 1.0

INTRODUCTION

With the recent advent of improved analytical and biomarker detection capabilities, a variety of organic chemicals have been found in trace amounts (Trace Organic Chemicals, TOrCs) in surface waters and fish tissue. TOrCs, also called contaminants of emerging concern, include pharmaceuticals, personal care products, surfactants, pesticides, flame retardants, and other largely unregulated organic chemicals. A subset of TOrCs, known as endocrine disrupting chemicals (EDCs), has received attention because of sex changes in fish (e.g., male fish developing female sex characteristics), and have been associated with effects on aquatic organisms in U.S. streams and rivers. Because TOrCs are widely used in nearly every sector of our society, many enter, and are discharged from, wastewater treatment plants (WWTPs). While wastewater treatment removes many, but not all, TOrCs, some are discharged into surface waters.

The growing awareness of TOrCs has lead utility operators to ask several questions including:

- Is my utility at risk of discharging TOrCs to the receiving stream?
- Which TOrCs, if any, contribute to observed effects on aquatic organisms?
- Which TOrCs should I sample in my effluent?
- Is there a way of identifying wastewater-influenced sites that are more likely to exhibit ecological effects due to TOrCs?
- Are the aquatic populations and communities in the receiving waterbody potentially at risk from TOrCs?

Effects of TOrCs are especially difficult to assess because they often co-occur with other, less subtle stressors, and because physiological or anatomical effects on individual organisms may or may not have population or community level effects. Most field information on TOrCs has addressed exposure but relatively little information is available linking exposure observations to population or community effects in aquatic systems. Identifying or predicting ecological effects of TOrCs in typical aquatic multi-stressor situations requires a variety of epidemiological tools that together can diagnose effects at multiple scales of ecological organization.

To help address these and other questions of the water resource community related to TOrCs, the Water Environment Research Foundation (WERF) is conducting a two-phase project. This report describes the Phase 1 research. The goal of this Phase 1 research was to determine the relationship between exposure to TOrCs and adverse impacts to aquatic biotic populations and communities. The primary objective of this Phase 1 research was to develop a screening tool that the water quality community can use to: 1) assess whether observed site-specific impacts on aquatic populations or communities are caused by TOrCs, and 2) determine whether exposure concentrations of TOrCs are likely to cause aquatic population and/or community effects. This research addressed five specific objectives:

(1) Develop and apply a procedure to prioritize which TOrCs are of most concern.

(2) Develop and test a conceptual site screening framework.

(3) Evaluate and test diagnostic approaches to identify potential risks due to TOrCs using various case studies.

(4) Develop a relational database and user interface with which the water resource community can enter, store, and search TOrC exposure data in the U.S.

(5) Foster partnerships and transfer knowledge gained in this research to the water quality community.

With over 500 organic compounds thus far detected in surface waters and effluent in the U.S., and well over 10,000 organic chemicals that are produced or used in the U.S., the universe of TOrCs is overwhelming. In order to provide meaningful information for the water resource community, while working within the time and funding allowed for Phase 1 of the research, TOrCs were defined as organic compounds that are known or suspected to be released to the aquatic environment, but are not commonly regulated or monitored, and whose potential risks to ecological health are relatively unknown. The focus of this Phase 1 research was therefore:

- Risks to aquatic populations and communities, not human health risks.
- Organic chemicals for which aquatic life criteria have not been developed and which are generally unregulated.
- Surface water sites and wastewater influenced sites (i.e; point sources) in particular.

The general framework addressed in this Phase 1 research is shown in Figure 1-1. Specific tasks performed in this research are summarized in Table 1-1. This framework, as well as specific screening and diagnostic tools that might be useful in this research, were discussed at a 2-day expert workshop convened in January 2010 (Diamond et al., 2010a). Suggestions and recommendations from that workshop were incorporated into our research strategy, which is discussed in three separate documents corresponding to different aspects of this research: 1) prioritization of TOrCs for ecological assessments (Diamond et al., 2010b), 2) screening and diagnostic tools for assessing risks of TOrCs (Diamond et al., 2010c), and 3) evaluation of diagnostic tools using several case studies (Diamond et al., 2010d).

This final report summarizes Phase I research including:

- The process used in this WERF Phase 1 project to prioritize TOrCs for further evaluation.
- A process to screen wastewater effluent for potential risk to aquatic organisms from TOrCs.
- An evaluation of the screening process and diagnostic tools using several case studies.
- A discussion of outstanding uncertainties and critical data gaps.

This report also discusses recommendations for Phase 2 research to further answer many of the questions identified in Phase 1. Phase 2 research will build on the partnerships and collaborative relationships established during Phase 1. Phase 2 research will focus on leveraging existing and proposed research on TOrCs, supplementing monitoring or data collection at established or proposed sites to increase the value of the information for assessing biological effects of TOrCs, and providing consistent, comparable, and compatible data for testing and refining both retrospective and prospective analytical tools and models.

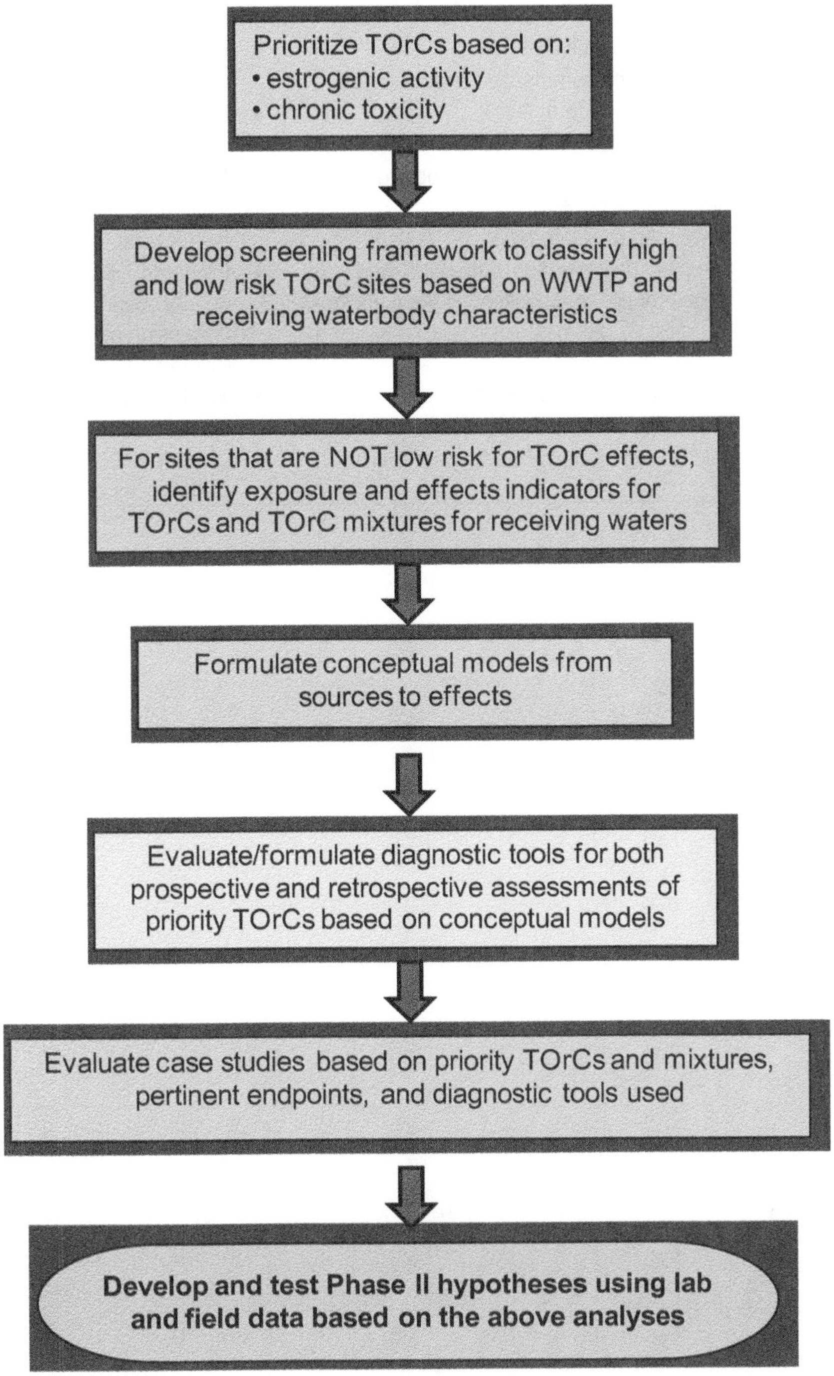

Figure 1-1. Flowchart Summarizing the Approach Used in Phase 1 Research Regarding Ecological Effects of Trace Organic Compounds (TOrCs).

Table 1-1. Summary of Specific Tasks Performed in Phase 1 Research.

Task Area	Description	
Prioritize TOrCs (Task Area 1)	• Critically evaluated prioritization approaches used	• Developed and verified priority list of TOrCs
Diagnostic Tools (Task Area 2)	• Critically evaluated diagnostic ecological tools and recommendations • Drafted risk hypotheses and conceptual models frameworks	• Identified preliminary analysis tools • Developed conceptual approach and screening tools
Case Studies (Task Area 3)	• Identified case studies • Compiled and summarized available data, metadata, and data gaps • Evaluated case study results	• Evaluated pros and cons of diagnostic tools used • Identified uncertainties, data gaps
Relational Database (Task Area 4)	• Refined relational database structure and metadata for project • Developed user interface	• Populated database with initial data • Developed documentation and user instructions
Recommendations (Task Area 5)	• Compiled uncertainties, data gaps, and recommendations identified in previous tasks	• Prepared Final report with a summary of results obtained in Tasks 1-3 and a discussion of uncertainties, data gaps, and recommendations for Phase 2.
Communication and Collaboration (Task Area 6)	• Developed Collaboration Plan • Developed Communication Plan • Held an expert workshop for peer review	• Identified funding and collaboration opportunities for Phase 2 • Developed fact sheets, presented research at conferences and workshops

CHAPTER 2.0

WHICH TORCS SHOULD I MONITOR? A FRAMEWORK FOR PRIORITIZING TRACE ORGANIC COMPOUNDS

Identifying or predicting ecological effects of TOrCs in typical aquatic systems is challenging, requiring prioritization of those TOrCs most likely to present a risk to aquatic ecological populations and communities so that monitoring and assessment efforts are properly focused. In the first task of this research (Diamond et al., 2010b) we assembled a database of TOrCs and developed three approaches to prioritize these compounds: 1) risk-based, 2) chemical persistence, bioaccumulation potential, and toxicity (PBT), and 3) a hybrid approach based on risk, persistence, and bioaccumulation potential. The following sections describe how data were collected for the database and how the prioritization was determined.

2.1 Development of TOrC Prioritzation Database

Over 100 publications were initially reviewed for the development of the TOrC database. Data appearing in the occurrence database were assembled from an in-depth review of 70 peer-reviewed journal articles (three of which were review articles) and reports (including a report by the International Joint Commission on emerging contaminants in the Great Lakes Basin; Klecka et al., 2009) that had established QA/QC protocols. Emphasis was placed on TOrC occurrence data from freshwater systems and effluents because monitoring data for these media are most readily available. Studies were sought that monitored primarily unregulated organic chemicals. In many cases, these studies also included more commonly monitored and regulated organic chemicals such as certain PCBs, PAHs, and legacy pesticides. The researchers included all organic chemicals monitored in a given study in the occurrence database. However, because they focused on studies of unregulated TOrCs, this occurrence database should not be construed as comprehensive regarding currently regulated chemicals. Many geographic areas in the U.S. and Canada are represented in this database and both effluent-dominated and non-effluent surface waters are represented.

Occurrence information alone is not sufficient for prioritizing TOrCs; occurrence concentrations must be placed in context of relevant ecological effects information as well as fate characteristics. Toxicity information for TOrCs was compiled from EPA's ECOSAR and estrogenic activity of TOrCs was compiled using the Food and Drug Administration's database. In addition, fate information was compiled for these TOrCs using EPA's EpiSuite and PBT profiler.

The majority of compounds found on the occurrence list could be grouped into one of 10 categories as listed below. The chemical categorization scheme used in our research and by other researchers is useful but imperfect as many of these chemicals are used in multiple ways; e.g., siloxanes were categorized as a personal care chemical (used widely in perfumes and soaps) but could have equally been categorized as an industrial chemical (defoamer). A total of 516 TOrCs were identified in our occurrence database, with the majority of TOrCs representing current use pesticides, pharmaceuticals, and industrial chemicals (Figure 2-1).

- Deodorizers/fragrances
- Flame retardants
- Industrial chemicals
- Natural hormones/steroids
- PAHs
- Personal care products
- Pesticides
- Pharmaceuticals
- Plasticizers
- Surfactants

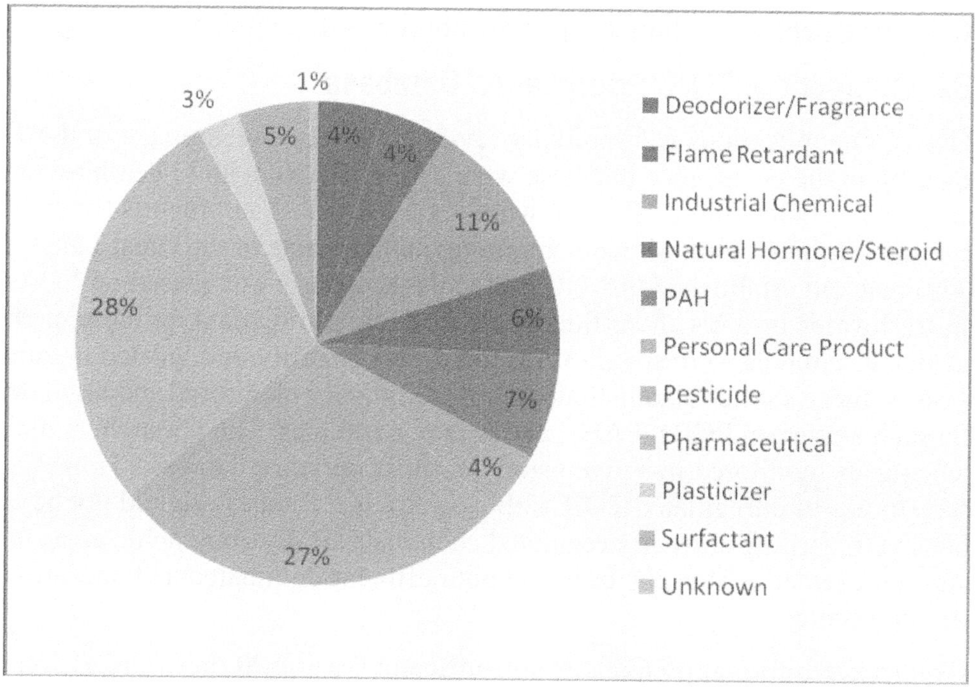

Figure 2-1. Grouping of TOrCs on the Occurrence List.

2.2 Determination of Risk

Risk was determined by integrating multiple types of information including occurrence and/or fate characteristics (bioaccumulation potential, persistence). Risk was calculated using three approaches for each TOrC for which both effects and occurrence data were available. First, a risk value (RV) based on the most sensitive predicted chronic toxicity endpoint (i.e., fish, *Daphnia*, or algae) was determined (see Equation 2.2-1). An RV greater than 1.0 indicates that the maximum observed concentration of the selected TOrC exceeds the most sensitive predicted toxicity threshold.

$$Risk\ Value = \frac{Max\ Occurrence\ Concentration}{Lowest\ Chronic\ Toxicity\ Threshold} \tag{2.2-1}$$

Two other risk quotients were calculated based on estrogenic effects using a system similar to that used for sediment quality guidelines. Concentrations of EE2 were identified that *are not* likely to have an effect on aquatic life (i.e., a predicted no effect concentration or PNEC) and those that *are* likely to have an impact on aquatic life (i.e., probable effect concentration or PEC). The PNEC for EE2 used was 0.35 ng/L (Caldwell et al., 2008) and the PEC used was 5 ng/L (Kidd et al., 2007). The estrogenic activity values of each compound were then multiplied by the PNEC and PEC to derive a PNEC and PEC for each TOrC with known estrogenic activity. Observed maximum occurrence concentrations and PNEC and PEC values were then used to calculate either a No Effect Endocrine Activity Risk Value (No Effect EARV) or a Probable Effect Endocrine Activity Risk Value (Probable Effect EARV) for each TOrC (where applicable) using Equations 2.2-2 and 2.2-3.

$$No\ Effect\ EARV = \frac{Max\ Occurrence\ Concentration}{Normalized\ PNEC} \tag{2.2-2}$$

$$Probable\ Effect\ EARV = \frac{Max\ Occurrence\ Concentration}{Normalized\ PEC} \tag{2.2-3}$$

2.3 Three Approaches Used to Determine TOrC Risk

Calculated risk quotients, toxicity and/or fate information was used to determine overall risk based on three approaches:

- **Approach 1**: The first approach is a simplistic but conservative risk-based prioritization based on 1) predicted chronic toxicity or estrogenic activity of a chemical (whichever is more sensitive) and 2) its maximum observed concentration in the occurrence database. While the occurrence data includes many studies and sites, we recognize that is not complete and may under-represent the potential maximum environmental concentration for some chemicals and sites. Maximum rather than median or mean observed concentrations were used for each TOrC because this provides a measure of conservatism. Also, maximum values were commonly reported in the literature whereas median or mean occurrence concentrations were less frequently presented. Additional inspection of several sources of occurrence data indicated that the maximum value reported was typically within an order of magnitude of the median value when sufficient data were available for a given TOrC. However, to help account for these uncertainties, a risk quotient of 0.1 (as opposed to 1.0) was used as a high priority threshold in this approach.

- **Approach 2:** The second approach uses a risk quotient similar to the first approach but converts the quotient to a score (1-3) that equates to low, medium, or high potential risk (Table 2-1). The approach also uses the same scale (1-3) to score bioaccumulation potential (based on log K_{OW}) and persistence (based on half-life in water) for each TOrC. The maximum score for the three factors is 9. High priority chemicals were those with scores a maximum of 3 for any one category or at least a total score of 7.

Table 2-1. Scoring Parameters Used in Approaches 2 and 3.

Parameter	Used in Approach #	Low Risk (1)	Intermediate Risk (2)	High Risk (3)
Max of Risk Value, NEERAV, or PEERAV	2	< 0.01	< 0.1 to ≥ 0.01	≥ 0.1
log K_{ow}	2 and 3	< 3.0	< 5.0 to ≥ 3.0	≥ 5.0
Half-life in water	2 and 3	< 60 days	< 180 to ≥ 60 days	≥ 180 days
Toxicity	3	> 100 µg/L	> 10 to ≤ 100 µg/L	≤ 10 µg/L
1/No Effect EAV	3	< 0.00001	< 0.01 to ≥ 0.00001	≥ 0.01
1/Probable Effect EAV	3	< 0.00001	< 0.01 to ≥ 0.00001	≥ 0.01

- **Approach 3:** This approach was similar to the "PBT" approach used by many organizations. It differs from the other two approaches in this research primarily because the method does not use occurrence concentration information or risk quotients based on occurrence data. TOrCs were scored in an approach similar to Approach 2 except toxicity scores are based on the predicted toxicity values obtained from ECOSAR and PBT profiler or the inverse of the no effect endocrine activity value (EAV) or the probable effect EAV. High priority chemicals were those which scored at least 7.

Each of the three prioritization approaches summarized above have advantages and limitations (see Table 2-2). For example, while Approach #1 is fairly simple to construct and communicate, and has reasonably high relevance related to ecological assessments, the prioritized list of TOrCs is dependent on TOrC occurrence data, which is always incomplete. In addition, several researchers have noted that there is significant variability and false positive/negative issues that may confound occurrence data, particularly when low level reporting limits are required for some TOrCs. Thus, the TOrCs considered, and the maximum concentrations recorded may not truly represent TOrCs of highest concern in surface waters. This limitation is true for Approach #2 as well. Therefore, it is important that users understand both the advantages and limitations of each prioritization approach presented.

2.4 Results of Risk Classifications

The compounds appearing on the prioritization lists along with chemical category, and some of the known modes and mechanisms of action are presented in Appendix A, Table A-1. Table 2-3 lists the high priority TOrCs identified using each approach. The compounds from the prioritization list are not intended to be implemented as a required monitoring list but instead, a starting point for reducing the universe of compounds found in surface and wastewater to make the task of TOrC screening more manageable. The three approaches differed in their identification of compounds with the potential to cause the most ecological risk to aquatic communities. The proportions of general categories of TOrCs were different depending on which approach was used (Figure 2-2). High priority TOrCs using Approach #1 consisted primarily of natural hormones and steroids, pharmaceuticals, and surfactants. This list contained most f the compounds that accounted for estrogenicity in a recent WWTP study (Furlong et al. 2010). Approach #2 yielded a high percentage of pesticides, and Approach #3 identified pesticides and industrial chemicals as the most prevalent groups of TOrCs.

Table 2-2. Advantages and Limitations of the Three Different Approaches Proposed for Prioritizing Trace Organic Chemicals (TOrCs).

Approach	Description	Advantages	Limitations
1	Risk-based: max concentration divided by most sensitive predicted effects threshold (toxicity or estrogenicity). Quotient ≥ 0.10 considered high priority.	Simple to use and communicate; based on actual measurements of TOrCs; considers TOrCs that are known to occur in surface waters; relatively high relevance to ecological risk.	Dependent on maximum concentrations observed and recorded in occurrence database; does not consider high production TOrCs not monitored.
2	Risk-based and fate-based: Sum of effect, bioaccumulation, and persistence scores; effect score based on quotient as in Approach #1; quotient ≥ 0.10 = score of 3 (highest priority); log K_{ow} ≥ 5.0 = score of 3; half-life in water ≥ 180 d = score of 3. Total score ≥ 7 is high priority TOrC.	Uses effects, bioaccumulation and persistence information for each TOrC-addresses fate as well as effects; based on TOrCs known to occur and that have been measured; relatively high relevance to ecological risk.	Dependent on TOrCs and their maximum concentrations in the occurrence database; does not consider HPV TOrCs not monitored.
3	Tox and fate-based: Sum of toxicity, bioaccumulation, and persistence scores; toxicity score based on predicted chronic toxicity; bioaccumulation and persistence scores same as in Approach #2.	Uses widely-accepted "PBT" approach-consistent chemical prioritization approaches used by other agencies; considers high production volume TOrCs as well as other TOrCs; not necessarily dependent on occurrence information; capable of identifying TOrCs that should be evaluated or have methods developed.	May include TOrCs that are unlikely to result in exposure to aquatic biota; based on toxicity effects, not endocrine effects; toxicity thresholds may have high uncertainty for some HPV TOrCs that don't have QSAR models.

It should be noted that the list of high priority TOrCs identified in this research is not identical to the list produced by the WERF TOrC treatability project CEC4R08 (*Trace Organic Compounds Removal during Wastewater Treatment Processes by their Efficacy in Reduction of a Suite of Indicator TOrC*). This is expected because the criteria used to prioritize TOrCs for evaluation in that project are necessarily different from the criteria used in this research. Whereas toxicity or endocrine activity was an important criterion determining high priority TOrCs in our research, toxicity was a relatively minor factor in the treatability research because other factors, such as occurrence (i.e., continuously present and of high enough concentration to measure removal across a process), physical/chemical characteristics (i.e., volatilization, adsorption onto solids, biotransformation), and analytical amenability and reliability (i.e., the ability to accurately analyze the TOrCs in raw sewage, treated wastewater, and biosolids at environmentally relevant concentrations) were more important from a treatability standpoint. None of these factors were important criteria in determining high priority TOrCs in our research; rather we relied on occurrence in terms of maximum concentrations reported in surface waters and effluents (not necessarily continuously), and physical/chemical characteristics in terms of TOrC log Kow (bioaccumulation potential) and half-life in water (a measure of aquatic persistence), not adsorption onto solids or volatilization per se. Finally, the prioritization approaches did not require that a reliable analytical method exists although to be measured as an occurrence, some analytical method was used. This issue is discussed later in this chapter.

Table 2-3. High Priority TOrCs Obtained from Each of the Three Prioritization Approaches.

Approach #1	Approach #2	Approach #3
Coprostanol	Pentachlorophenol	3-methylcholanthrene
17α-ethynylestradiol	PBDE-209	o,p'-DDE
Cholesterol	Tonalide	Mirex
Stigmastanol	Galaxolide	p,p'-DDE
β-sitosterol	Tamoxifen	cis-Nonachlor
Stigmasterol	Hexabromocyclododecane	p,p'-DDT
17β-estradiol	Desmosterol	p,p'-DDD
Bis(2-ethylhexyl)adipate	Cholesterol	o,p'-DDT
PBDE-209	Epicoprostanol	trans-Nonachlor
Bis(2-ethylhexyl)phthalate	Coprostanol	γ-Chlordane
Mestranol	Campesterol	trans-Chlordane
Di-N-octyl phthalate	Stigmasterol	Trifluralin
Desmosterol	β-sitosterol	Ethalfluralin
Cholestanol	Stigmastanol	Benfluralin
4-Nonylphenol	cis-Chlordane	PBDE-209
4-n-nonylphenol	Bis(2-ethylhexyl)adipate	Miconazole
Estrone	4-nonylphenol diethoxycarboxylate	cis-Chlordane
Testosterone	4-nonylphenol monoethoxylate	pentachloro-Benzenethiol
Novobiocin	4-Nonylphenol	3-chloro-2,6-dinitro-N,N-dipropyl-4-(trifluoromethyl)-Benzenamine
Campesterol	4-n-nonylphenol	Aldrin
4-nonylphenol diethoxycarboxylate	4-nonylphenol monoethoxycarboxylate	Bicyclo 2.2.1 hept-5-ene-2,3-dicarboxylic acid, 1,4,5,6,7,7-hexachloro-, dibutyl ester
4-nonylphenol monoethoxycarboxylate	Bis(2-ethylhexyl)phthalate	Lofepramine
17α-estradiol	Di-N-octyl phthalate	Octachlorostyrene
Bisphenol A	para-nonylphenol	1,3,6,8-tetrabromo-Pyrene
4-nonylphenol monoethoxylate	17α-ethynylestradiol	2,4,6-trimethyl-2,4,6-tris(3,3,3-trifluoropropyl)-Cyclotrisiloxane
Epicoprostanol	Oryzalin	1,1'- 1,2-ethanediylbis(oxy) bis 2,4,6-tribromo-Benzene
para-nonylphenol	Musk ketone	tetracosafluorotetradecahydro-Phenanthrene
Galaxolide	Mestranol	1H-Isoindole-1,3(2H)-dione, 2,2'-(1,2-ethanediyl)bis 4,5,6,7-tetrabromo-
Norethisterone	Triclosan	1,4:7,10-Dimethanodibenzo a,e cyclooctene, 1,2,3,4,7,8,9,10,13,13,14,14-dodecachloro-1,4,4a,5,6,6a,7,10,10a,11,12,12a-

Table 2-3. Continued.

Approach #1	Approach #2	Approach #3
Tamoxifen	Acetyl cedrene	Dibenz[a,h]anthracene
Tonalide	Fluoranthene	trans-permethrin
Estriol	OTNE	Heptachlor
Propranolol	Celestolide (ADBI)	Endrin
Fenoprofen	Benzo[a]pyrene	o,p'-DDD
Oryzalin	Clotrimazole	Hexachlorobenzene
Equilenin	Traseolide (ATII)	Oxychlordane
Hexabromocyclododecane	Musk xylene	cis-permethrin
Musk ketone	Methoxychlor	Pentachloroanisole (PCA)
3-tert-butyl-4-hydroxy anisole	Ethalfluralin	Methoxychlor
Pentachlorophenol	Benfluralin	Perylene
Triclosan	Trifluralin	Benzo[g,h,i]perylene
	Endrin	Permethrin
	Dieldrin	Indeno[1,2,3-c,d]pyrene
	Pentachloroanisole (PCA)	Pendimethalin
	Heptachlor	Benzo[e]pyrene
	Oxychlordane	Benzo[b]fluoranthene
	Hexachlorobenzene	Benzo[k]fluoranthene
	o,p'-DDD	Epicoprostanol
	o,p'-DDE	Phantolide (AHMI)
	p,p'-DDD	Hexabromocyclododecane
	trans-Nonachlor	Celestolide (ADBI)
	cis-Nonachlor	Campesterol
	trans-Chlordane	Traseolide (ATII)
	Miconazole	Musk xylene
	3-methylcholanthrene	Tamoxifen
	p,p'-DDE	Dieldrin
	o,p'-DDT	Benzo[a]pyrene
	Mirex	Desmosterol
	p,p'-DDT	Stigmasterol
	γ-Chlordane	Pentachlorophenol
		β-sitosterol
		Tonalide
		Stigmastanol
		Galaxolide
		Cholesterol
		Coprostanol
		Acetyl-hexamethyl-tetrahydro-napthalene (AHTM)
		Cyclohexasiloxane, dodecamethyl-Cyclohexasiloxane
		2,4,6-tris(1,1-dimethylethyl)-Phenol
		Perthane
		2-(2H-benzotriazol-2-yl)-4,6-bis(1,1-dimethylpropyl)-Phenol
		Hexadecamethyl heptasiloxane
		octadecamethyl-Octasiloxane
		Ergosterol
		tetrabromo-, bis(2-ethylhexyl) ester Phthalic acid

Table 2-3. Continued.

Approach #1	Approach #2	Approach #3
		Heptachlor epoxide
		2-Methylfluoranthene
		4-n-octylphenol
		Benzo[b]naphtho[2,1-d]thiophene
		Propargite
		Benzo[a]anthracene
		Chrysene
		Clotrimazole
		Chlorpyrifos
		Mestranol
		4-tert-octylphenol
		Musk ketone
		17α-ethynylestradiol
		Fluoranthene
		Acetyl cedrene
		OTNE
		Di-N-octyl phthalate
		para-nonylphenol
		Bis(2-ethylhexyl)adipate
		Bis(2-ethylhexyl)phthalate
		4-n-nonylphenol
		4-Nonylphenol
		Octylphenol
		β-estradiol-3-benzoate
		Heptamethylphenylcyclotetrasiloxane
		1,1'-ethylidenebis 3,4-dimethyl-Benzene
		2,4,6,8-tetraethenyl-2,4,6,8-tetramethyl-Cyclotetrasiloxane
		1,1,1,5,5,5-Hexamethyl-3,3-bis[(trimethylsilyl) oxy]-trisiloxane
		Dodecylmethylpentasiloxane
		triphenyl ester Phosphorous acid
		1,1'-Biphenyl, bis(1-methylethyl)-
		1,1,1,5,5,5-hexamethyl-3-phenyl-3-[(trimethylsilyl)oxy]-Trisiloxane
		1,3-DiNO2 2-CL 5-CF3 benzene

The risk-based approach resulted in the fewest number of high priority TOrCs (41) while the PBT approach resulted in the greatest number of high priority TOrCs (108). A total of 23 TOrCs were identified as high priority using all three approaches and a total of 126 TOrCs were identified as high priority using any one of the approaches (see Table A-1, Appendix A). For most utilities and resource agencies beginning the process of examining TOrCs, the risk-based approach may be most appropriate because these TOrCs are known to occur in surface waters in the U.S., they are thought to elicit effects on aquatic biota at relatively low concentrations, and they appear to sometimes occur at concentrations that might result in a risk to aquatic biota.

Results of the prioritization analyses indicated that many pharmaceuticals currently being monitored are unlikely to occur at concentrations that would present a risk to aquatic biota based on predicted toxicological or estrogenic activity endpoints. Exceptions were synthetic hormones and steroids. In addition, the risk-based approach suggested that predicted chronic toxicity endpoints were more sensitive than endpoints based on estrogenic activity for most TOrCs. Again, the exception was hormones and steroids. The TOrCs for which estrogenic activity thresholds were more sensitive (and therefore used instead of toxicity thresholds in the risk quotient) included: 17α-estradiol, E2, EE2, bisphenol A, estriol, estrone, equilenin, mestranol, methoxychlor, norethisterone, tamoxifen, and testosterone. These results may not be surprising in light of the fact that most TOrCs have purportedly weak estrogenic activity, if any, but they may be chronically toxic at relatively low concentrations in comparison with maximum occurrence concentrations identified in this research. However, the estrogenic activity levels for many TOrCs is presently uncertain and other types of endocrine disruption effects of TOrCs could not be addressed based on the current state of knowledge.

The difference in TOrCs identified as high priority using the three approaches is exemplified by the distribution of chemical persistence categories observed. Nearly half of the high priority TOrCs using risk-based Approach #1 had relatively low persistence in aquatic systems (half-life < 60d) while 85% of TOrCs using Approaches #2 and #3 (in which persistence is a prioritization factor) had half-lives > 60d and 35-45% had half-lives > 180d (Figure 2-3). The relatively high percentage of non-persistent TOrCs using the risk-based approach, coupled with the possibility that some of these TOrCs may be frequently, or even continuously supplied to waterbodies (e.g., via treated wastewater discharges), suggests that chemical persistence may not always be a useful indicator of TOrC risk in aquatic systems.

The lists of high priority TOrCs generated in this research should not be taken as required monitoring lists or as compound lists for regulation because much is uncertain regarding the occurrence and effects of many TOrCs given our current state of knowledge. High priority TOrCs are expected to vary somewhat with the region of the U.S. and types of TOrCs that may be generated. However, the prioritization approaches presented in this report should prove useful for a utility or locality to help them organize, reduce, and manage the process of screening TOrCs.

All three approaches used to identify high priority or high risk TOrCs in this research have similarities to the methodologies used by various water resource agencies to assess risk of more commonly monitored chemicals (e.g., metals, legacy pesticides) in that a chemical by chemical approach is used. However, for TOrCs of interest in this research, the environmental protectiveness of this type of approach is questionable. Many monitoring studies have documented that a number of TOrCs may occur together from a single source or at a site (Kolpin et al., 2002; Daughton and Teine, 1999; Snyder et al., 2003) and several laboratory studies have demonstrated additive effects of certain types of TOrCs (e.g., estrogenic hormones: Kortenkamp and Altenberger, 1999; Brian et al., 2007). These results indicate that while a chemical by chemical approach may be acceptable for developing screening risk quotients and prioritizing TOrCs that pose the greatest risk, actual site assessments and diagnostic tools need to consider the cumulative risk of the mixture of TOrCs present.

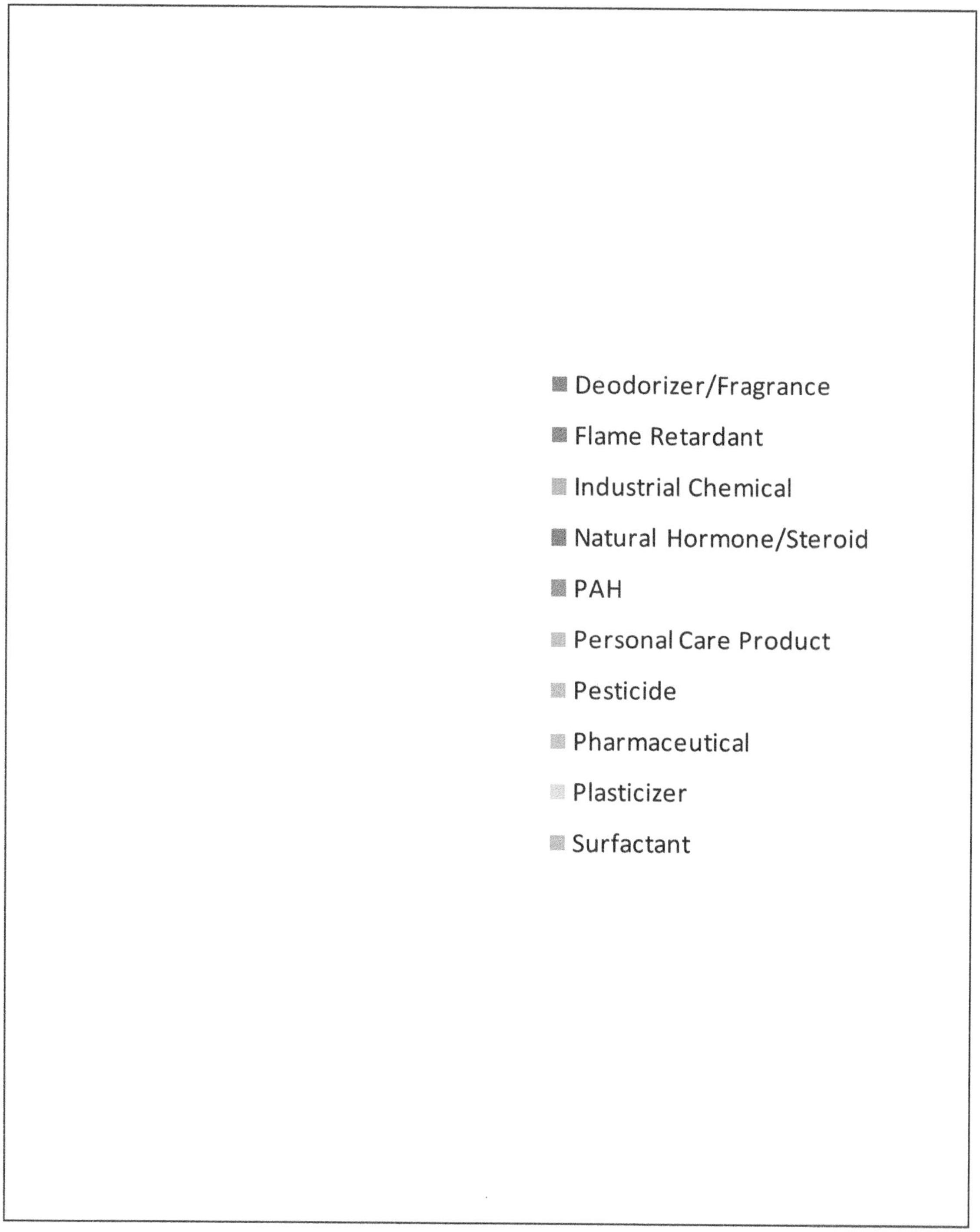

Figure 2-2. Percentage of Different Types of High Priority TOrCs Identified Using Each of the Three Prioritization Approaches. "Risk" approach is based on the most sensitive endpoint divided by the maximum concentration observed for each TOrC; P = persistence (half-life in water) expressed as a score between 1 (half-life < 60 d) and 3 (half life > 180 d); B = bioaccumulation potential (log Kow) expressed as a score between 1 (Log Kow < 3) and 3 (Log Kow > 5). T = toxicity expressed as a score between 1 (most sensitive chronic value > 100 ug/L) and 3 (chronic value < 10 ug/L).

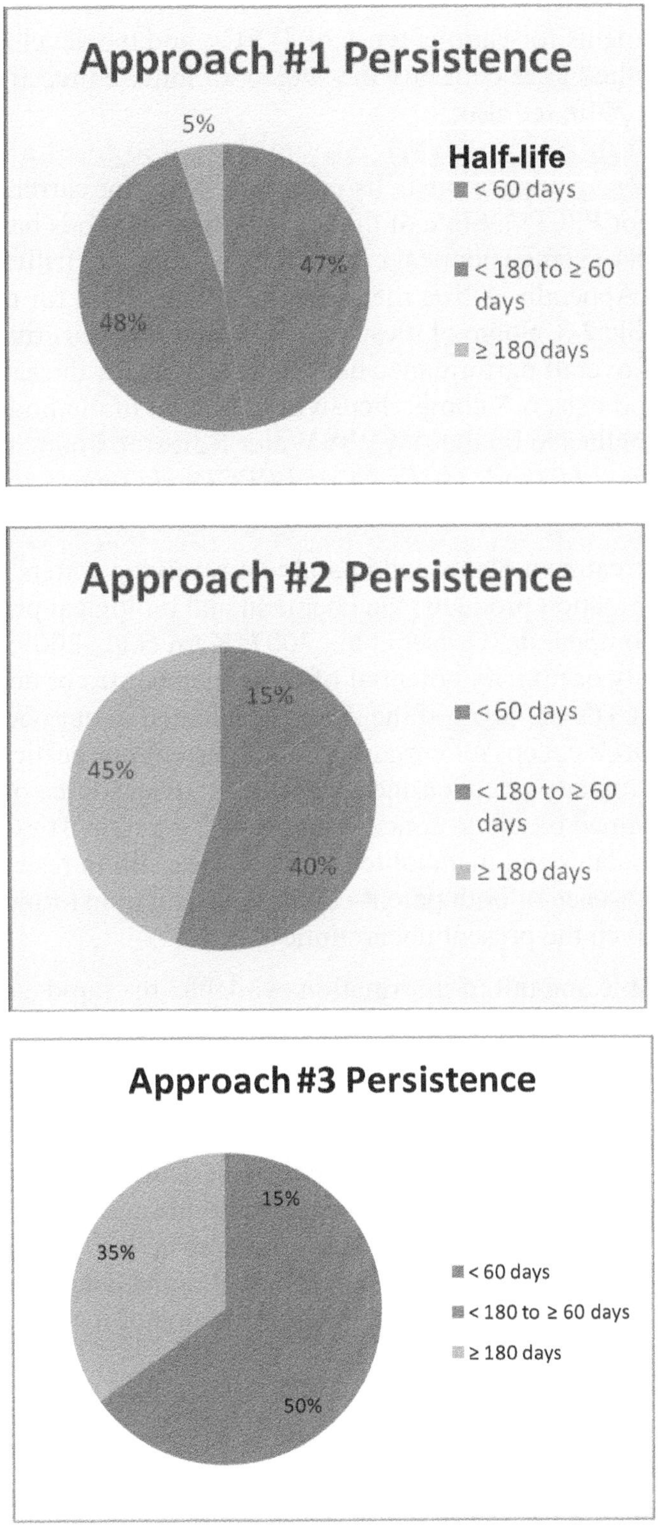

Figure 2-3. Percentage of TOrCs Within Each Persistence (half-life) Category Resulting from Each Prioritization Approach.

2.5 Sampling and Analytical Methods Assessment

Analytical measurements for various types of TOrCs, and the development of new sampling and analytical methods, are currently the subject of intense investigation (e.g., see Hannah et al. 2009; Ort et al. 2010; see also http://toxics.usgs.gov/highlights/pharmaceuticals_method.html and http://water.epa.gov/scitech/swguidance/methods/ppcp/index.cfm for current status of EPA and USGS analytical methods for PPCPs). State of the art analytical methods have permitted the detection and quantification of some chemical constituents at parts per trillion (ppt) concentrations. Table A-2, Appendix A lists methods that are available for most of the high priority TOrCs listed in Table 2-3. Some of these methods are more sensitive than others, and there are different levels of overall performance between methods for the same substances, where more than one method exists. A comprehensive evaluation of methods for some listed TOrCs is currently being conducted by the AWWA Water Research Foundation (WRF Project 4167: *Evaluation of Analytical Methods for EDCs and PPCPs via Inter-Laboratory Comparison*).

Within wastewater treatment plants and in ambient receiving waters, reactions of parent compounds to yield transformation products, via chemical and biological processes, are known to occur readily for many compounds (Gasser et al., 3007; Kern et al., 2009; Helbling et al., 2010). The endocrine activity or toxicity potential of these metabolites or degradates are currently unknown for most TOrCs. Most of the studies evaluated in our occurrence database did not report degradates of TOrCs except for certain types of current use pesticides. Thus, it is possible that the long-term toxicity or endocrine activity of a given source of TOrCs may be greater or less than that assumed based on concentrations of the parent TOrCs. In most cases, analytical methods for degradates and metabolites of TOrCs are still in research. It may be important to evaluate the presence of both parent compounds and transformation products in future research of TOrCs given the present uncertainties.

Given the considerable amount of information available, the rapid expansion of the current knowledge base, and the variable performance of measurement methods, categories of emerging contaminants should be addressed as groups, rather than individual substances. This approach is being considered by U.S. EPAs' drinking water program in order to be more efficient and keep pace with increasing knowledge about chemicals (http://blog.epa.gov/dwstrat/). Measurement methods exist to analyze multiple analytes (see Table A-2), often with relatively small additional cost. In addition, given some of the assumptions made in prioritizing compounds in this research, it seems reasonable to assume that certain additional compounds, including byproducts, are, or will be become important. Additional data on a broader list of compounds can be obtained relatively easily and, no doubt, some of them will be important, as additional information becomes available.

CHAPTER 3.0

What Types of Wastewater-Influenced Sites Are at Most Risk Due to TOrCs?

In this part of the research, the research team developed a framework and tools for use by the water quality community to screen sites for TOrC risks to aquatic biota (Diamond et al., 2010c). If it can be demonstrated that a site is low risk for TOrC effects, extensive monitoring and assessment of TOrCs is probably not warranted. If the site could have a high risk (or unknown risk) due to TOrCs, monitoring may be warranted. The TOrCs listed as high priority (Table 2-3) in the previous chapter could serve as a guide as to what type of TOrCs to monitor. This framework could be used both prospectively and retrospectively depending on the objective of the user.

3.1 Proposed Site Screening Approach

Based on the availability of data and current knowledge, the preferred site screening framework is to identify sites with the highest potential risk of impacts based on known ecological information (e.g., effluent dominated, limited mixing, or closed system, etc.) and determine which of these actually exhibit biological impairment due to the presence of TOrCs. This process has the advantage of relying on relatively accessible information that, in many cases can be obtained remotely through public databases. There are many potential sources of TOrCs to aquatic systems, with treated wastewater discharges being only one, albeit important source type. Therefore, site screening, to help prioritize areas most at risk from TOrC effects, is essential. The screening approach developed in this research focuses on wastewater discharge sources and is a first step toward developing an effective screening tool. Figure 3-1 and Table 3-1 present hypothesized risk factors used in the screening process and characteristics are grouped as: 1) wastewater influent characteristics, 2) wastewater treatment characteristics, 3) receiving system characteristics, and 4) site ecological and TOrC concentration data (measured or predicted). Using the guidance from information gathered for Table 3-1 one can further categorize sites to determine if a specific site warrants further investigation for TOrC impairments. Sites are likely to range from low risk to high risk and can be categorized as shown in Table 3-2.

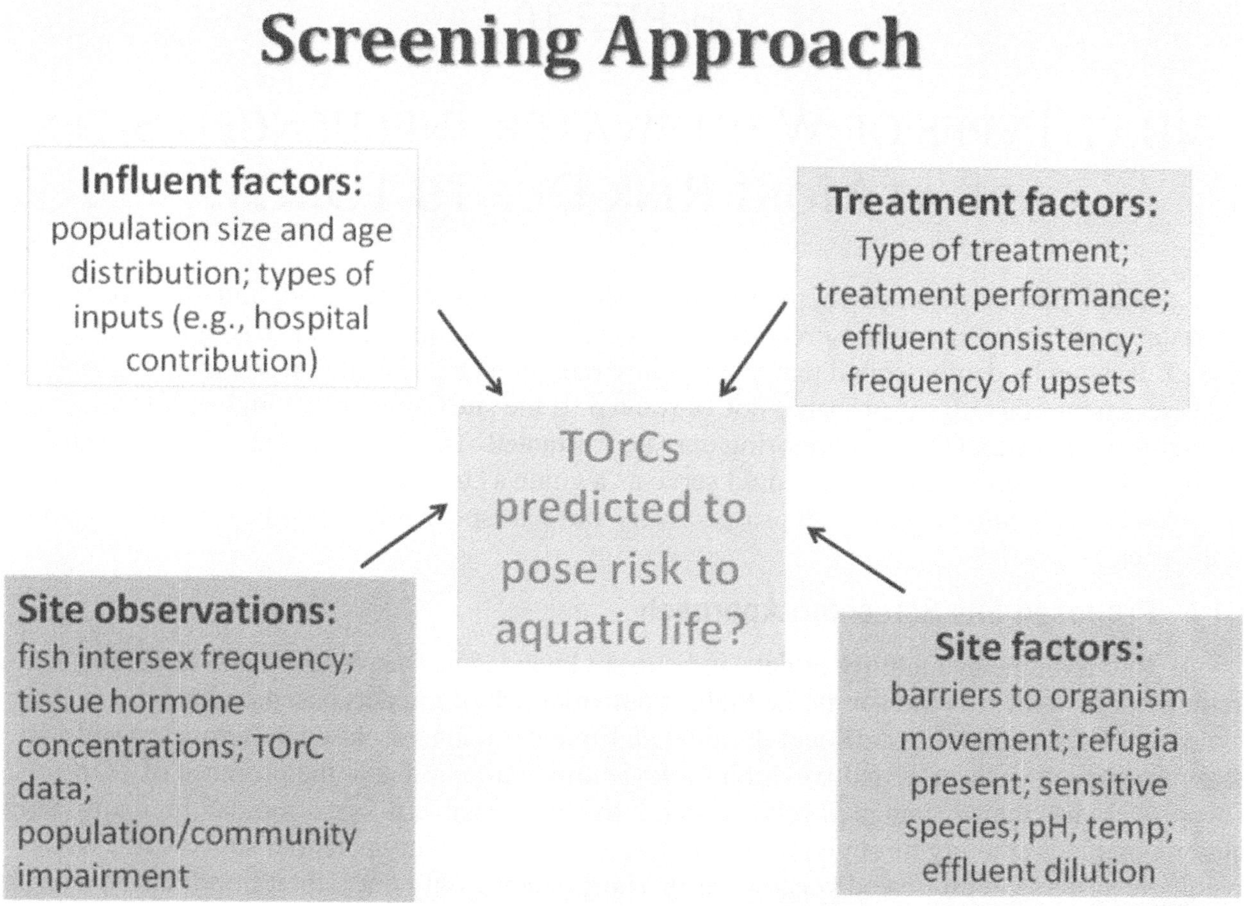

Figure 3-1. General Schematic Showing Hypothesized Factors Considered in Screening Wastewater Sites for Diagnostic Analyses Involving TOrCs.

Table 3-1. Draft Criteria for Screening Waste Water Treatment Plant (WWTP) Effluent-Influenced Sites for Potential Aquatic Ecological Effects Due to Trace Organic Chemicals.

	Factor	Risk Thresholds: Lower Risk	Risk Thresholds: Higher Risk
WWTP Input Characteristics	Industry/hospital-related influents	None or very minor	>5% of influent
	Population served	< 10,000	> 1M
	Average age of population	< 30	> 50
WWTP Treatment Characteristics	Level of treatment	Tertiary, advanced secondary	Secondary or Primary
	Sludge retention time (SRT)	> 15d	< 5d
	Treatment effectiveness/uniformity	Low (near reference condition) nutrient concentrations; TSS	High (relative to reference conditions) nutrient levels; TSS
	WWTP upsets	None	Several
	Uniformity of influent flow and water quality	Relatively uniform most of the time	Very inconsistent in terms of flow and/or quality
Receiving waterbody	Percent Effluent Instream (low flow)	< 1%	> 70%
	Waterbody size and/or habitat diversity	Large (e.g., > 5th order) waterbody; abundant refuges	Small waterbody; refuges few or non-existent
	Waterbody openness/barriers	No barriers to fish movement	Dams or other barriers preventing fish movement
	Threatened, endangered, or species of concern present	None	At least 1
	Presence of other potential sources of TOrCs	None	Agricultural (CAFOs, row crops, orchards) sources present upstream or nearby urban stormwater?
Site Observations	Existing EDC effects present	None	Significantly higher vtg induction in male fish; abnormal frequency of intersex in fish
	TOrC concentration > screening threshold level (from literature values)	No	Yes

Table 3-2. Summary of Different Risk Categorizations Based on Screening Criteria.

TOrC RISK	Site Risk classification	WWTP Facility	Receiving Waterbody
↓	**LEVEL I** (Lowest)	Low risk	Low risk
↓	**LEVEL II**	Low risk	High risk
↓	**LEVEL III**	High risk	Low risk
↓	**LEVEL IV** (Highest)	High risk	High risk

3.2 Types of Risk Scenarios for Screening Wastewater Influent Sites

Based on observations from Tables 3-1 and 3-2, and other site specific data, screened sites likely fit into one of the following overall scenarios. (Table 3-3):

A) *Possible Risk:* TOrCs are likely to be present at levels of concern, but it is not known whether they are causing or could cause population or community level effects

B) *Possible Risk*: Receiving environment impairments have been observed, but actual TOrC exposure is unknown or incompletely known so it is uncertain whether TOrCs or other stressors are actually responsible for observed community impacts

C) *Unknown Risk:* It is not known whether TOrCs occur at levels of concern or whether there have been receiving water effects.

D) *Low Risk*: Existing studies indicate no ecological effects due to TOrCs, and/or TOrCs either singly or in combination, are not above known levels of concern

A fifth, obvious scenario is when TOrCs are known to occur at levels of concern and there is biological impairment demonstrated at the site. This scenario would clearly be high priority for risk.

Table 3-3. Summary of Major Types of Scenarios to be Addressed in Screening Wastewater-Influenced Sites for Potential TOrC Risk to Aquatic Biota.
If TOrCs are known to exceed thresholds and impact is observed, risk potential is high and further screening is probably unecessary

WWTP Effluent TOrC Scenario	TOrCs in the Effluent Exceed Conservative Thresholds?	Population/Community Impact Observed?	Risk Potential
A	Yes	Unknown	Possible
B	Unknown	Yes	Possible
C	Unknown	Unknown	Unknown
D	No	No	Low

3.3 Using the Screening Approach

The screening approach was applied with modifications in a retrospective assessment of available data from wastewater-influenced sites in the Drift Plain ecoregion of Ohio (Diamond et al., 2010c). This assessment used a GIS approach to integrate hypothesized screening criteria in this research with modeled TOrC exposure concentrations and quantitative stressor-response associations generated from screening-level statistical analyses. The method integrated the GIS work with results from the consumer products industry model, iSTREEM (American Cleaning Institute), to identify sites that included a WWTP and impaired biological conditions downstream of the effluent discharge. The method accounted for other potential stressors including physical, chemical, habitat, and biological parameters that could influence biological condition of the aquatic environment. Impairments were judged based on the Ohio EPA fish IBI index and relative fish weight (a ratio of actual fish weight to what a healthy fish of the same weight and length should weigh) as compared to reference condition values for the same ecoregion. Site risk was based on risk scores for WWTPs and general risk of water bodies receiving effluent, similar to the screening approach shown in Figure 3-1. A total of ten candidate sites were selected based on risk to fish community condition, and seven were selected for risk relative to fish weight. From those, two were chosen as case studies – Ravenna WWTP, a discharge to Wahoo Ditch (a tributary to Breakneck Creek, Cuyahoga River) and Mansfield WWTP, a discharge to Rocky Fork (a tributary to Black Fork of the Mohican River). The site screening tool predicted that the significant biological stressors in Wahoo Ditch included personal care products and pharmaceuticals (PPCPs) from the WWTP, along with septic influences and human population density. Impairment in Rocky Fork was also predicted due to WWTP PPCPs. Other stressors at this site included high conductivity and human population density. These two sites were further assessed using EPA's CADDIS framework as described in the next chapter.

In addition to the two Ohio sites discussed above, five other case studies were examined with which the researchers tested the screening framework, as well as diagnostic tools as described in Chapter 4.0 (Table 3-4). The case studies represented a range of scenarios and breadth of monitoring information available that are likely to encompass the types of situations most utilities and water resource managers are likely to face.

All of the case study sites, with the exception of Southern California Bight (SCB) and Guelph were classified as higher risk and warrant further examination (Table 3-4). The SCB and Guelph sites were classified as lower risk potential because of either high effluent dilution and no barriers to organism movement (SCB) or good wastewater treatment and few other TOrCs input sources (Guelph). To date, both Guelph and SCB have not produced evidence suggesting TOrC effects at the population or community level; i.e., organismal effects such as intersex or lesions appear to be absent and available bioassessment data (hornyhead turbot for SCB and both macroinvertebrate and wild fish for Guelph) indicate little or no impairment. For the remaining five case study sites, the proposed screening approach indicated a higher potential for risk due to TOrCs, largely due to the type of wastewater treatment available (secondary treatment with little or no nutrient removal, presence of potential TOrCs input sources in addition to domestic sources) and relatively high effluent concentration under low flow conditions.

The criteria used in this screening approach are in essence a set of working hypotheses that should be tested further in Phase 2 research, using information gained from related projects such as WERF Project CEC4R08: "Trace Organic Compounds Removal during Wastewater Treatment Processes by their Efficacy in Reduction of a Suite of Indicator TOrC". The interactions between wastewater influent characteristics, types of wastewater treatment, and ecological factors are very complex and not well understood at this time. For example, the screening tool places higher TOrC risk for WWTPs that have substantial influent flow from hospitals and other facilities in which pharmaceutical and personal care products are likely. A recent study by Phillips et al (2010) suggests that WWTPs receiving wastewater from pharmaceutical manufacturers may be at greater risk of having pharmaceutical TOrCs. While this finding is not inconsistent with the screening tool, it underscores the fact that there are as yet many unknowns in terms of sources of TOrCs to WWTPs. Specific hypotheses developed during this project should help define a roadmap of needed collaborative research that will refine the screening approach so as to provide a valuable tool for better water resource management.

Table 3-4. Summary of Case Study Sites Examined.

Site (WWTP)	Location	Percent Effluent[1]	Potential Stressors in Addition to WWTP TOrCs	Biological Impairment?[2]	TOrC Data?	Scenario Type[3]	Risk Level[4]
SCB[5]	CA	<1	Legacy pesticides and PCBs; normal ocean temperature variation	Not Known	Water, Fish, Effluent, Sediment	A	I or III, depending on WWTP
Guelph	ON	40	Channelization; low-head dams; agricultural runoff	Not known	Water, Fish, Effluent	A	II
Kitchener	ON	10 - 15	Urban runoff; low-head dams; agricultural runoff; CAFOs[6]	Yes, fish	Water, Fish, Effluent	A	IV
Ravenna	OH	>90	Urban runoff; septic systems	Yes – fish, macroinvertebrates	Predicted	B	IV
Mansfield	OH	>80	Urban runoff; legacy industrial contaminants; industrial discharges; agricultural runoff	Yes – fish, macroinvertebrates; but upstream impaired too	Predicted	B	III
Taylor Run	PA	85	Urban runoff, stormwater	Yes – periphyton, macroinvertebrates	Fish Tissue	A	III
Boulder	CO	>70	Urban runoff; legacy mining pollutants; agricultural runoff	Not Known	Water, Fish, Effluent	A	Pre-upgrade= IV Post upgrade = II

[1] Under low flow or summer conditions

[2] Population or community level effects (e.g., changes in species abundance, biomass, age classes) observed downstream of the WWTP?

[3] Scenario categories based on Task 2 research in this project: A = TOrCs exceed conservative effect thresholds and instream effects have not been evaluated; B = Unknown whether TOrCs exceed thresholds and effects have been observed instream

[4] Risk levels based on screening tool developed in Task 2 of this project: Level I = low TOrC risk to aquatic biota in terms of wastewater treatment plant (WWTP) and instream factors; Level II = low risk based on WWTP factors but high risk based on instream factors; Level III = high risk based on WWTP factors but low risk based on instream factors; Level IV = high risk based on both WWTP and instream factors.

[5] Southern California Bight - Represents 4 WWTPs in and around Los Angeles

[6] Concentrated Animal Feeding Operation

CHAPTER 4.0

DIAGNOSING POTENTIAL OR ACTUAL ECOLOGICAL EFFECTS DUE TO TORCS

Diagnosing population or community-level effects of TOrCs in aquatic systems is as, or more challenging than defining the screening tool. TOrCs comprise a variety of chemicals with different MOAs, different spatial and temporal distributions depending on their sources, and to a large extent, unknown or uncertain effects on aquatic organisms. For example, while many TOrCs are believed to be capable of producing endocrine disrupting effects in some types of organisms, estimates of actual endocrine activity potential are often lacking for these chemicals. Similarly, toxicity data for many TOrCs is based on QSAR information, which may or may not accurately capture this form of risk for some TOrCs. In addition, TOrCs are unlikely to occur in the absence of other potential stressors; e.g., hydrological modifications, nutrient enrichment, and conventional toxics, as well as natural pressures (floods). The multiple stressor aspect of this issue makes the development of diagnostic tools in this research very challenging.

TOrC effects can be diagnosed prospectively or retrospectively. A conceptualization of prospective versus retrospective diagnostic assessments of TOrCs evaluating exposure in relation to biological effects is shown in Figure 4-1. This approach relies on combining TOrCs with similar modes of action to estimate true TOrC exposure and potential effects. To evaluate effects, a retrospective assessment would use the approach described in EPA's CADDIS and/or Canada's EEM framework in addition to one of several analysis tools listed in Diamond et al. (2010c). Prospective assessments would need to rely on lab or field-derived thresholds by TOrC type data (e.g., EDCs) to provide an estimate of risk in terms of different levels of biological organization. Many of the same tools used to diagnose observed effects in retrospective assessments can be used prospectively as well, such as species sensitivity distributions (Posthuma et al., 2002) and models to predict population/community effects. Not all criteria will be relevant to all stressor types at a site due to the types of data available, how much is known about stressor-response relationships and the degree to which MOA is known for a given stressor. For example, distinguishing nutrient effects would require either a reference condition analysis approach (i.e., nutrient ecoregions) (U.S. EPA, 2000b) or water quality model (e.g., Qual2K), which could be used for either prospective or retrospective assessments, or an approach that recognizes differences in species tolerance to nutrients, or functional ecosystem effects that are diagnostic in a retrospective assessment. The case studies discussed in Diamond et al. (2010d) are primarily retrospective analyses as these are helpful to understand the challenges of prospective studies.

The research team compiled information for many diagnostic approaches and tools used to identify cause of biological impairments in aquatic systems (Diamond et al., 2010c), some of which are summarized in Table 4-1. Several challenges become important when attempting to diagnose whether TOrCs are a cause of observed impairment or could be a risk when conducting prospective assessments. First, at higher levels of organization (population, community), there may not be real "diagnostic" impacts that are attributable only to specific TOrCs (or classes of TOrCs). A decline in fish growth rate, for example, could be due to declines in food resources, as much as to impaired physiology associated with chronic TOrC exposure. Therefore, the diagnostic approach must use multiple lines of evidence and look for patterns of responses that are not attributable to other stressors. Second, both exposure and effects indicators are needed to diagnose whether particular TOrCs are causing, did cause, or could cause population or community-level ecological effects at specific sites. Third, for many TOrCs, particularly those that act on aquatic communities either through bioaccumulative pathways or through endocrine disrupting mechanisms, population or community effects may or may not be observed pending many other site factors.

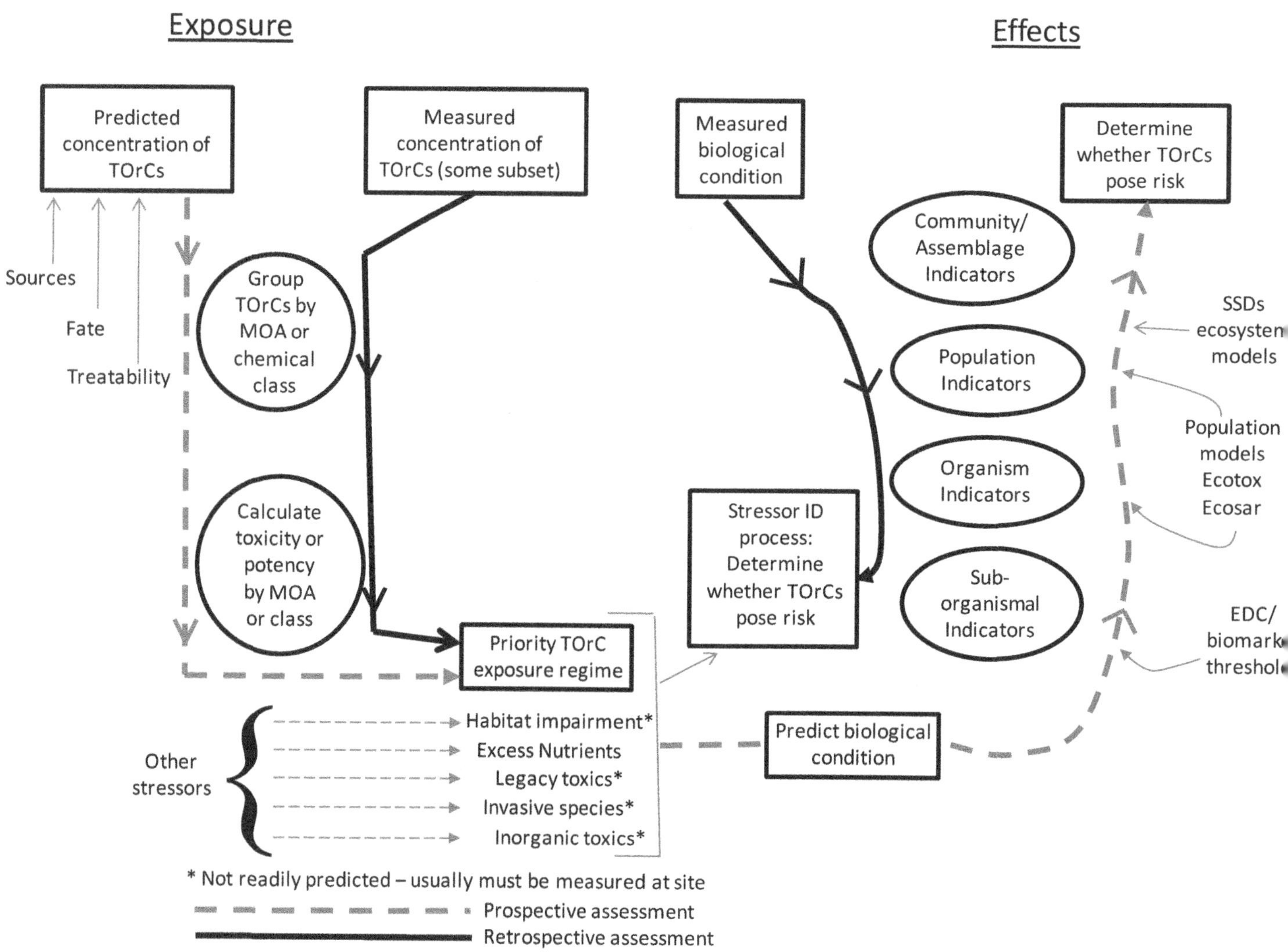

Figure 4-1. Flow Diagram Conceptualizing Steps Needed in Prospective and Retrospective Assessments of Trace Organic Chemicals (TOrCs) in Terms of Ecological Impact.

Table 4-1. Summary of Potentially Useful Diagnostic Tools That Could be Used to Identify Whether TOrCs or Other Stressors Cause Observed Impacts. Stressors, Resources and Tools Listed are Examples; Others May be Appropriate for a Particular Site. Sources for the Stressors Would Need to be Identified for Each Particular Case. SSD - Species Sensitivity Distribution Analysis; ICM - Interspecies Correlation Models; CA & RA - Concentration and Response Addition Models; QSAR – Quantitative Structural Activity Relationship Models

Stressors		B-H Criteria[1]	Diagnostic Approach[2]	Resources/ Analyses	Analysis Tools
-BOD/ DO stress		SA; C; P	Ref; TS	DO taxa sensitivity data; reference DO regime; DO effects database	DO models; fish habitat suitability models (HSM)
-Temperature stress		SA; C; P;M	TS; AS	Temp taxa sensitivity data; reference regime; 316(a) database	Fish HSMs; conditional probability
-Excess nutrients		SA; C; P	Ref; TS; AS	Ecoregional nutrient database; algal sensitivity data; algal metrics; invertebrate metrics	WQ models (e.g., Qual2E); conditional probability; ecological models
-Hydrological modification		T; C; M	Ref; TS	Taxa flow requirements;	HSMs; GIS models; population models
-Sedimentation		SA; C	Ref; TS; AS	Sediment taxa sensitivity data; reference regime; invertebrate metrics	HSMs; conditional probability; population models
-Inorganic toxics	metals TDS ammonia	SA, C	Ref; TS; AS	Tox databases (e.g., ECOTOX); taxa sensitivity metrics; ref regime	SSDs; ICMs; CA&RA
-Toxic TOrCs	Pesticides PAHs LAS	SA; C; P	TS; AS; D	ECOTOX; taxa sensitivity metrics; ref regime; OPPT database	SSDs; ICMs; CA&RA
-"Other" TOrCs	PPCPs Organometals	SA; C; P	D; EM	PhACTS database; P&G data;	PhATE model; pop. and ecological models

1 – Bradford-Hill diagnostic criteria: T – temporality; SA – strength of association; C – consistency; P – plausibility; M – manipulation, evidence of recovery or lack of effect

2 – Ref – comparison with appropriate background, historic information, or reference condition regime; TS – consistency with taxa sensitivity information; AS – assemblage specificity information; D – diagnostic information (e.g., physiological, anatomical effects indicative of specific stressor); EM – ecological and population modeling

4.1 What is Meant by Ecological Effects?

A key element in diagnostic tool development, regardless of the approach used, is what constitutes an ecological effect. Based on discussions with several stakeholder representatives, **this research defines ecological impairment as a significant change in growth, reproductive potential, or abundance of a species (which, in its extreme may be observed as a change in species composition) as compared to the known natural baseline condition for that species, water body type, and geographic region.** The team recognizes that there are other types of effects that may or may not result in population/community impairment as defined above. Similar to the process considered for prioritizing TOrCs in Task 1, we recognize a gradient of effects that vary in terms of ecological relevance as noted by many other researchers. These include: 1) physiological effects (enzyme, hormone, other biomarker changes that signify exposure to certain stressors); 2) morphological effects (liver or gonad size change, intersex, changes in growth or fecundity); and 3) higher level effects (e.g., changes in population abundance) (Figure 4-2). Bradbury et al. (2004) discuss the need to link toxicological responses across different levels of biological organization. Compared to biochemical endpoints, community endpoints have high ecological relevance, but a long time lag, low reversibility, and present challenges tying changes back to a specific cause. At the other end of the spectrum, biochemical responses have a short time lag, high reversibility, and good potential for linkage with causes, but have relatively low ecological relevance. Some countries have developed monitoring programs that use indicators at all levels of organizations (i.e., Sweden's Environmental Effects Monitoring program; see Thoresson, 1996; Sandstrom and Neuman, 2003). While it is the population/ community effect that we are most interested in from a risk assessment perspective, we are also interested in making links between effects at lower levels and those at higher levels. That is perhaps the greatest challenge in developing the diagnostic tool.

4.2 Potential Indicators Diagnostic of Certain TOrCs

As in any type of ecological risk assessment involving chemical pollutants, measures of exposure can range from: (a) actual (e.g. analytical methods) or modeled (e.g. P*h*ATE, iSTREEM) chemical concentrations at a site, (b) physiological and other sub-organismal measures (i.e., biomarkers) that indicate organism exposure to a certain type of chemical (e.g., stress protein levels, Vtg., Table 4-2), and (c) morphological measures such as gonad-somatic index, liver size, and condition factor. Whole organism measures (such as deformities, lesions, and gill damage) or community-level effects (such as IBI or individual metrics such as percent mayflies), are not generally diagnostic of chemicals such as TOrCs (Suter et al., 2002), or even types of TOrCs, because by definition they occur at low concentrations in aquatic systems and do not generally elicit gross morphological effects at those concentrations (although they may indirectly cause such effects). Therefore, for diagnostic purposes, measures of TOrC exposure in this research rely primarily on TOrC concentrations at a site, as well as physiological and morphological measures that are known to be diagnostic of certain types of TOrC exposure.

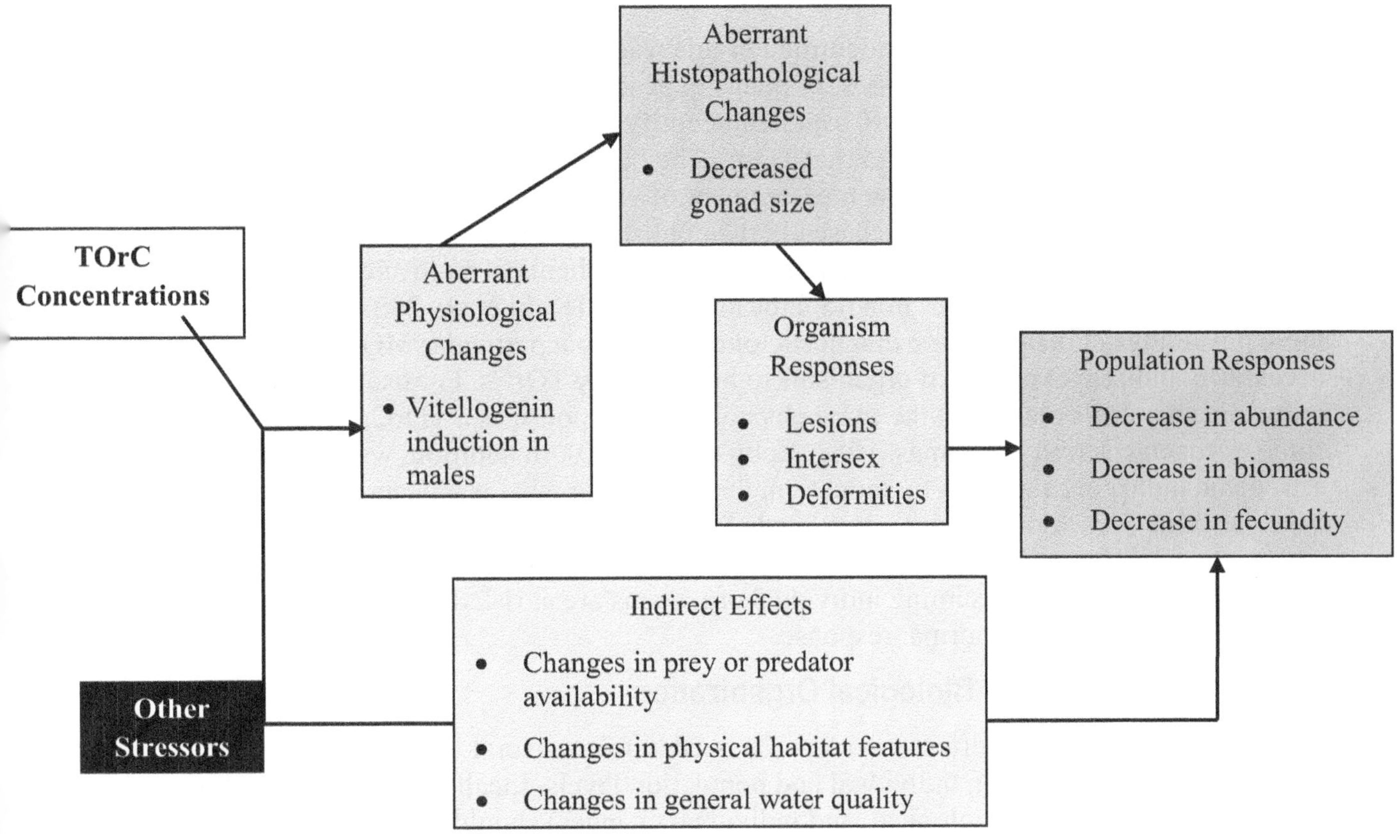

Figure 4-2. Relationships between TOrC Exposure and Exposure to Other Stressors and Different Levels of Effect.

Table 4-2. Examples of Physiological Measures That Have Been Used to Document Exposure of Fish and Other Biota to Various Trace Organic Chemicals.

Physiological Measure of Exposure	Applicable Type of Organism	Example TOrCs
Vitellogenin (fish) or Vitellogenin-like protein (mussel hemolymph)	Fish, mussels	Estrogenic compounds (natural/synthetic hormones, bisphenol A, nonylphenols, pesticides, cadmium, triclosan)
Hormone Measurements		
Estrogen, Progesterone	Fish	Estrogenic compounds (natural/synthetic hormones, bisphenol A, nonylphenols, pesticides, cadmium, triclosan)
Testosterone	Fish	Androgenic compounds (natural/synthetic hormones,
Thyroid hormones	Fish	PBDEs, PCBs, pesticides, cadmium
Nitro musk adducts in hemoglobin	Fish	Musk xylene
Acetylcholinesterase activity	Invertebrates, Fish	Organophosphate and carbamate pesticides
EROD induction (liver)	Fish	PAHs, other chemicals
Cytochrome P_{450} Activity	Invertebrates, Fish	A variety of chemicals. Activity or abundance of different P_{450} enzymes can be associated with specific chemicals.
Analytical measurements of chemicals in bile, blood, liver, or other tissues	Vertebrates or large invertebrates	Exposure to specific compounds could be determined.

Many high priority TOrCs identified in this research elicit specific physiological responses in vertebrates, invertebrates, or plants depending on their mode of action (MOA). Knowing the MOA for a given TOrC can help identify specific physiological measures that indicate exposure of the organism to that TOrC. The best-known example of this approach regarding high priority TOrCs is the measurement of vitellogenin (Vtg.) in male fish, which has been extensively described and discussed in the published literature (e.g., see Pait and Nelson, 2002 for discussion of Vtg. as an indicator of estrogenic chemical exposure) and was shown to be linked to population declines in wild fathead minnows (Kidd et al., 2007). While many of these physiological measures are diagnostic of exposure to a particular MOA, they may not necessarily indicate exposure of organisms to high priority TOrCs. For example, certain inorganic chemicals may elicit the same physiological response as a TOrC; e.g., lead may have some estrogenic activity (Thomas, 1988; Ruby et al., 2000). In addition, with respect to estrogenic indicators such as Vtg. induction, understanding what is an appropriate reference or baseline value (i.e., "normal" organism levels for the time of year, species, etc. in the absence of anthropogenic TOrCs) is critical to the interpretation of results. An inappropriate baseline can result in false positives: assuming individual organisms are at risk due to exposure of estrogenic TOrCs when in fact it is a normal response.

4.3 Linking Levels of Biological Organization

To diagnose TOrC effects, evidence is needed from several levels of biological organization; at a minimum, individual and population levels. Ideally, lower level biological responses (molecular, physiological, and individual changes) should be linked to higher level responses (population and community changes). A pivotal point in the diagnostic system for this research is the individual organism. A number of physiological or histopathological measures may provide evidence that an individual organism has been exposed to certain TOrCs but there still needs to be a link made between individual exposure, or even exposure of multiple individuals, and measurable population-level effects. While certain types of TOrCs may have effects only on one assemblage (e.g., estrogenic TOrCs affect vertebrates such as fish primarily), obtaining measures of effect for different assemblages is desirable to help diagnose whether stressors other than TOrCs may be responsible for observed effects. TOrCs may have both direct and indirect effects on a given species at a site, requiring different diagnostic tools. In addition, rarely will TOrCs be the only potential stressors at a site. This is particularly true for wastewater-influenced sites, which are often associated with human-modified systems. The seven case studies evaluated in this research demonstrate the types of stressors that often occur: urban runoff, CSOs, legacy sediment contaminants, hydrological modifications, and altered riparian zone characteristics (Table 4-3). All of these stressors could mask population-level effects caused by TOrCs. Therefore, it is necessary to use a defensible method of eliminating other potential stressors such as EPA's CADDIS or Canada's EEM methodologies.

The researchers evaluated several diagnostic approaches using the seven case studies mentioned in Chapter 3.0. Types of approaches included: EPA's causal analysis framework, CADDIS; Canada's Environmental Effects Monitoring (EEM) causal analysis approach; application of the American Cleaning Institute's iSTREEM model and the pharmaceutical manufacturer's P*h*ATE model to predict concentrations of certain TOrCs at a wastewater influenced site; and ecological modeling (Comprehensive Aquatic Systems Model [CASM]). Some of the case studies represent a prospective assessment while others represent a retrospective assessment in terms of determining causes of biological impairment at the site; i.e., evidence of biological impairment at the population or community level at some sites has not been observed and various techniques are illustrated to determine prospectively whether it is likely that TOrCs could have such effects in the future.

The evaluation of seven case studies in this research indicated that most focused on one or two biological levels of organization (Table 4-4). As a result few cause-effect linkages are present to connect individual changes or suborganismal alterations to population-level effects. An example illustrating the lack of data on linkages is illustrated in a conceptual model constructed for the Boulder Creek case study (Figure 4-3). The many light colored pathways in this example illustrate unknown or uncertain linkages due to lack of data. A similar trend is observed in the other case study conceptual models. Strongest linkages between TOrC exposure and individual-level effects were observed at the Kitchener and Boulder Creek sites, where both TOrC exposure data and diagnostic organismal exposure information (e.g., Vtg. levels in male fish, high intersex frequency) were available and consistent with estrogenic TOrCs having some effect at lower levels of biological organization (Table 4-4). Both of these sites were predicted to be high risk for TOrC effects and biological information appeared to support potential TOrC effects at some level of biological organization (Table 4-3).

Linkages to TOrCs at the remaining sites could not be determined due to lack of appropriate data (e.g., no organismal or suborganismal information was available; Table 4-4). The lack of diagnostic TOrC exposure data (i.e., suborganismal or organismal measures indicative of exposure to a certain class of TOrCs), or organism effects data (intersex, lesions etc.) at the two Ohio sites, for example, restricted our ability to diagnose whether certain TOrCs could have contributed to observed biological impairment at these sites. Similarly, while fish tissue data collected at the Taylor Run site indicated TOrC exposure, other types of potentially diagnostic organismal effects data (e.g., intersex frequency, lesions) were not available with which to identify likely linkages with TOrCs or other potential stressors (e.g., excess nutrients). Finally, the lack of population/community data available at the Boulder Creek site (at least pre-WWTP upgrade) limited our ability to identify whether TOrC or organism exposure data collected there was actually linked to any population or community-level effect. The results of the CASM model for the Boulder Creek site, in which only three of the many TOrCs measured there were evaluated, suggested that population level effects were possible at this site with only a slight increase (doubling) in TOrC concentration. These results underscore the need to have exposure and effects data at multiple biological scales to adequately diagnose whether TOrCs contribute to observed population/community impairment.

4.4 Summary

Several important conclusions were demonstrated in attempting to diagnose causes of observed impairment in the case studies, and specifically, determining whether TOrCs were a cause of impairment. First, a site-specific conceptual model with potential stressor sources, candidate causes, causal pathways, and ecological effects (both direct and indirect) is useful to help document hypotheses, highlight critical data gaps, indentify other potential stressors, and determine potential effects of TOrCs. Such models are routinely used in ecological risk assessments (U.S. EPA, 1998) and are useful in causal analysis as well (Suter et al., 2002).

Second, frameworks for organizing stressor identification information (EPA's CADDIS, Canada's Environmental Effects Management approach) are useful to present lines of evidence that support or rule out potential stressors in multi-stressor situations. These and other tools used in our case studies each have their strengths and limitations (see Table 4-5). Community-level bioassessments, for example, are not sufficient to indicate TOrC effects. However, robust data regarding fish, macroinvertebrate, and periphyton can be used with other tools to link and diagnose potential biological effects of TOrCs. In combination, several of these tools can be useful in diagnosing whether TOrCs are or could be a cause of biological impairment.

Third, a well-coordinated sampling and analysis approach is critical to link potential causes of biological impairment to population level effects. Sites should be sampled with attention to temporal and spatial patterns of biological endpoints such as intersex frequency and Vtg response in male fish. Exposure and effects information at multiple biological levels (minimally individual organism and population level) is critical to diagnosing TOrC effects. While TOrC concentration data in water, tissue, or sediment are helpful for identifying that exposure has occurred, suborganismal indicators (e.g. vitellogenin induction) are more useful for determining the effects of exposure.

Fourth, effluent-dominated sites are the best candidate sites for further refining and understanding the effects of TOrCs. Large waterbodies, such as the SCB case study site evaluated, have potentially complex fate and transport processes and TOrCs are more likely to co-occur with a variety of natural and anthropogenic factors that could mask the effects of TOrCs. Studies should be conducted at multiple sites to determine potential patterns and diagnostic characteristics of TOrCs and understand the impact of other types of stressors.

4.5 What Data are Needed to Improve the Effectiveness of Processes for TOrCs?

As pointed out in Chapter 4.0 and in Diamond et al. (2010d), most of the case studies lacked data for at least one level of biological organization that impeded our ability to evaluate causality between TOrCs and observed biological impacts. The following example, using the West Chester, PA case study, shows how certain types of additional analyses could increase the effectiveness of any causal analysis approach such as CADDIS or EEM in diagnosing whether TOrCs can be eliminated as a candidate cause at a site. Some of the analyses suggested could be conducted by a utility now depending on results of applying the screening tool. Further recommendations for filling data gaps and addressing unknowns are discussed in Chapter 6.0.

4.5.1 Summary of Current Site Knowledge

The West Chester site had fish, macroinvertebrate, and periphyton assessments up and downstream of the WWTP discharge point. Some impairment of the periphyton and macroinvertebrate assemblages was observed downstream of the discharge as compared to either upstream or an ecoregional reference site. The fish community was apparently unaffected by the effluent in terms of several widely used metrics. Habitat assessments indicated little or no impairment downstream and TOrC data was limited to a one time analysis of fish tissue data (species) downstream of the discharge only. Effluent pollutant data are available but do not include high priority TOrCs. Based on effluent data and a site reconnaissance, other candidate causes besides TOrCs at this site include excess nutrients and toxics and sedimentation from the adjacent interstate highway.

4.5.2 Critical Data Gaps and Suggested Analyses

(1) Data Gap: Effluent TOrC data are not available. The screening tool suggests that TOrCs are likely in the effluent given the type of treatment (trickling filter, chlorination) and the fairly high contribution of hospital waste as well as residential flow to the WWTP.

Suggested Analyses: Analyze effluent for at least a subset of the high priority TOrCs identified using Approach #1 (risk-based) as well as those detected in fish tissue. If possible, analyze effluent on two separate occasions, perhaps in relation to different seasons when removal efficiencies may be different. Analyze upstream samples as well to determine the background TOrC levels at the site.

(2) Data Gap: Fish metrics suggested no impact at this site but these metrics may not be sensitive to TOrC effects. TOrC exposure/effects data at the individual organism level are lacking. Given that fish tissues have accumulated certain TOrCs at the site, fish are clearly being exposed to at least those TOrCs.

Suggested Analyses: Analyze fish for intersex up and downstream of the discharge. Preferably examine more than one species to obtain representative data. Analyze vitellogenin levels in male fish. Check fish for lesions and other symptoms that are indicative of either chemical pollutants or pathogen effects.

(3) Data Gap: Nutrients are a potential stressor at this site. Macroinvertebrate impacts may be due to indirect effects of nutrients and not TOrCs. However, linkages between nutrients, algae, and observed macroinvertebrate impairment have not been firmly established thus far. While general fish metrics do not indicate impact, more subtle effects information regarding overall fish health and gonad condition could provide diagnostic information as to whether nutrients or TOrCs could be responsible for observed effects.

Suggested Analyses: Analyze macroinvertebrate species data with respect to nutrient sensitivity using nutrient-macroinvertebrate species distribution data collected by EPA and Pennsylvania Department of Environmental Protection for this region of Pennsylvania. If sensitive species are present in proportions expected based on reference conditions, nutrients are unlikely to be a cause of observed macroinvertebrate impacts. Analyze algal biomass and productivity at the site to determine food availability for macroinvertebrate and herbivorous fish species at the site. Analyze fish condition, liver size, and gonad size as per Canada's EEM approach. If condition, and liver and gonad size are all higher in fish from the site as compared to the reference condition, nutrient enrichment is a likely cause of impairments observed (see Figure 4-3). If, however, condition and liver size are higher but gonad size is lower than reference, this suggests that some form of metabolic disruption, such as from TOrCs, may be a cause of observed impacts (Figure 4-4; Table 4-6).

Table 4-3. Summary of Results for the Seven Case Studies Sites and Comparison with Screening Results.

Site	Screening Approach Risk Potential		Risk Level[3]	Observed Exposure/Effects	TOrCs Potential Cause of Impairment?	Other Potential Stressors Present
	WWTP[2]	Receiving System				
SCB[1]	L-H	L	I – II	Little evidence of TOrC exposure or effects in horny head Turbot population	Unlikely for this population	Temperature-ocean currents; legacy pollutants
Guelph	L	H	II	Little or no population or community effects observed	Unlikely	Channelization
Kitchener	H	H	IV	Decrease in abundance of several fish species; decrease in fish species diversity; intersex increase in darters	Potentially (high intersex rate in rainbow darters) but other stressors appear to be a more significant source of impairment	Ammonia, nutrients, habitat modification
Ravenna	H	H	IV	Impairment of fish and macroinvertebrate communities.	Potentially, but other stressors appear to be a significant source of impairment	Nutrients, habitat, septic system leakage, urban runoff, ammonia
Mansfield	H	L	III	Impairment of fish and macroinvertebrate communities.	Potentially—other stressors may be responsible for impairment	Contaminated sediments (metals, PCBs, pesticides), urban runoff
Taylor Run	M	M-H	III	Fish tissue concentrations of pharmaceuticals and personal care products	Potentially but unlikely.	Urban runoff, agriculture
Boulder Creek	H	H	IV	Elevated concentrations of estrogens downstream of WWTP and high rate of intersex/ uneven sex ratio in white suckers	Potentially—high rate of intersex /uneven sex ratio in white suckers but population endpoints lacking	Urban runoff, agriculture, historical mining

1 SCB – Southern California Bight – 4 large WWTP discharges
2 L = low, M = medium, H = high risk potential
3 Level of potential ecological risk due to TOrCs; I = lowest risk, IV = highest risk

Table 4-4. Summary of Available Information for Each of the Seven Case Studies, Relevant to Different Levels of Biological Organization and TOrC Diagnostic Analyses.

Site	Suborganism	Organism	Population	Community
SCB[1]	Vtg.[2] in male turbot; TOrCs and legacy CECs in livers; cortisol and thyroid hormones in blood	Ova-testis; intersex rate	Gender ratio; relative abundance	ND[3]
Guelph	TOrC analyses *in water*; Vtg., *hormone*, enzymes in fish exposed in situ	14-d *in situ* fish studies	Fish *population* studies	Macroinvertebrate and fish bioassessments
Kitchener	TOrC analyses in water; Vtg., hormone, enzymes in fish exposed in situ	Intersex rate;	Darter population studies	Macroinvertebrate and fish bioassessments
Ravenna	ND (predicted TOrC risk)	ND	ND	Macroinvertebrate and fish bioassessments
Mansfield	ND (predicted TOrC risk)	ND	ND	Macroinvertebrate and fish bioassessments
Taylor Run	TOrCs observed in fish tissue	ND	ND	Algal, macroinvertebrate, and fish bioassessments
Boulder Creek	Vtg. in male white suckers	Intersex in white suckers	Skewed sex ratio in white suckers	ND

1 SCB – Southern California Bight
2 Vtg. – vitellogenin analysis
3 ND – No data

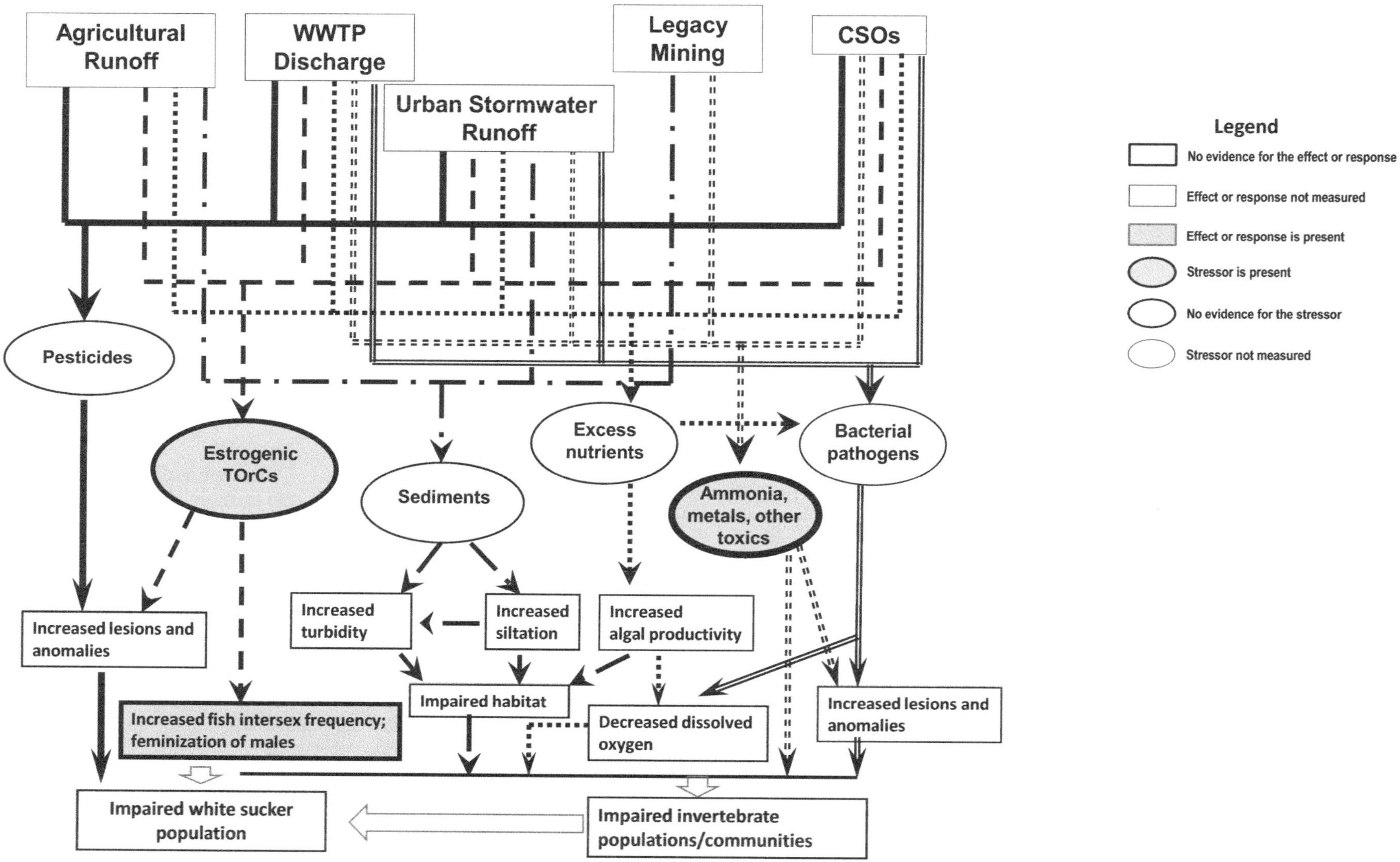

Figure 4-3. Conceptual Model for Boulder Creek Case Study Site Illustrating the Current Uncertainty in Many Pathways and Stressors Due to Lack Of Data.
(Note: Bold lines indicate pathway exists, light colored lines indicate pathway was not examined.)

Table 4-5. Summary of Data Requirements, Strengths, and Limitations of Different Diagnostic Tools and Relative Cost.

Tool or Approach	Data Requirements	Stengths	Limitations	Relative Cost[1]
CADDIS[2]	• Robust bioassessment data at multiple sites	• Good way to organize available information and systematically evaluate the data	• Retrospective approach, not predictive	$$
	• Stressor data that is co-located with biological data	• Capable of integrating different lines of evidence	• Unlikely to provide clear conclusion regarding potential TOrC effects unless diagnostic analyses are available as well as evidence at multiple steps in the causal pathway.	
		• May be useful approach for eliminating non-TOrC candidate causes	• Often is unable to definitively eliminate other candidate causes because analyses predominantly rely on bioassessment data, which in itself is rarely diagnostic of a single stressor.	
		• Uses bioassessment data available from State monitoring programs.		
EEM[3]	• Effects data at multiple biological scales	• Tiered or phased approach – more sophisticated analyses are done as needed	• Data intensive at higher tiers	$$$
	• Biological data at multiple sites typically and/or different time periods	• Capable of making linkages between different levels of biological organization	• Could take years to complete if a large or complex site	
	• Stressor data that is co-located with biological data	• May be useful for eliminating non-TOrC candidate causes	• Retrospective approach, not predictive	
		• Capable of integrating different lines of evidence		

Table 4-5. Continued.

Tool or Approach	Data Requirements	Advantages	Disadvantages	Relative Cost
Predictive TOrC Exposure Models (e.g., iSTREEM, PhATE)	• Land cover, land uses	• Potentially useful tool to characterize TOrC exposure at a site	• Uncertainties regarding accuracy of predictions not well characterized	$
	• WWTP treatment type	• Does not require sampling or travel to site	• Does not currently address TOrCs of most concern at a site	
	• TOrC fate information	• Required information can be obtained via accessible sources	• Considers toxicity not estrogenic activity (iSTREEM)	
	• Population estimate	• Capable of characterizing cumulative risk potential of TOrCs (iSTREEM)	• Requires extensive high quality bioassessment information, including habitat quality info (iSTREEM)	
		• Capable of ranking relative strength of stressors including TOrCs at a site (iSTREEM)	• Does not address stressors that are either unrelated or indirectly related to land uses or are modified due to site-specific conditions (iSTREEM)	
			• Relies on community-level data to represent biological responses – does not incorporate diagnostic biological info. (iSTREEM).	
			• Relies on predicted TOrC concentration to differentiate high from low TOrC risk sites (iSTREEM)	
Ecosystem Models (e.g., CASM)	• TOrC concentration – response data	• Potentially capable of characterizing importance of indirect toxicity-based effects of TOrCs	• Requires sophisticated model analysis and baseline information regarding a site or area.	$$
	• Habitat/system autecological information (e.g., temperature regime, water depth)	• Useful tool for establishing critical testable hypotheses for future research	• Currently limited to those TOrCs with well-characterized concentration-response data	
	• Dissolved nutrient concentrations	• Potentially useful tool for predicting population risk due to TOrC toxicity	• Currently limited to toxicity-based responses; estrogenic potential data has not been modeled.	
		• Possible to address cumulative impacts of TOrCs with similar modes of action	• Autecological and nutrient data may need to be estimated because site data are unavailable.	
		• Possible to characterize relative risk of TOrCs versus other stressors of concern at a site (e.g., nutrients).		

[1] $ = Relatively less costly $$ = more costly; $$$ = most costly
[2] CADDIS = Causal Analysis/Diagnosis Decision Information System (cfpub.epa.gov/caddis).
[3] EEM = Environmental Effects Monitoring (www.ec.gc.ca/esee-eem/default.asp?lang=En&n=4CDB9968-1).

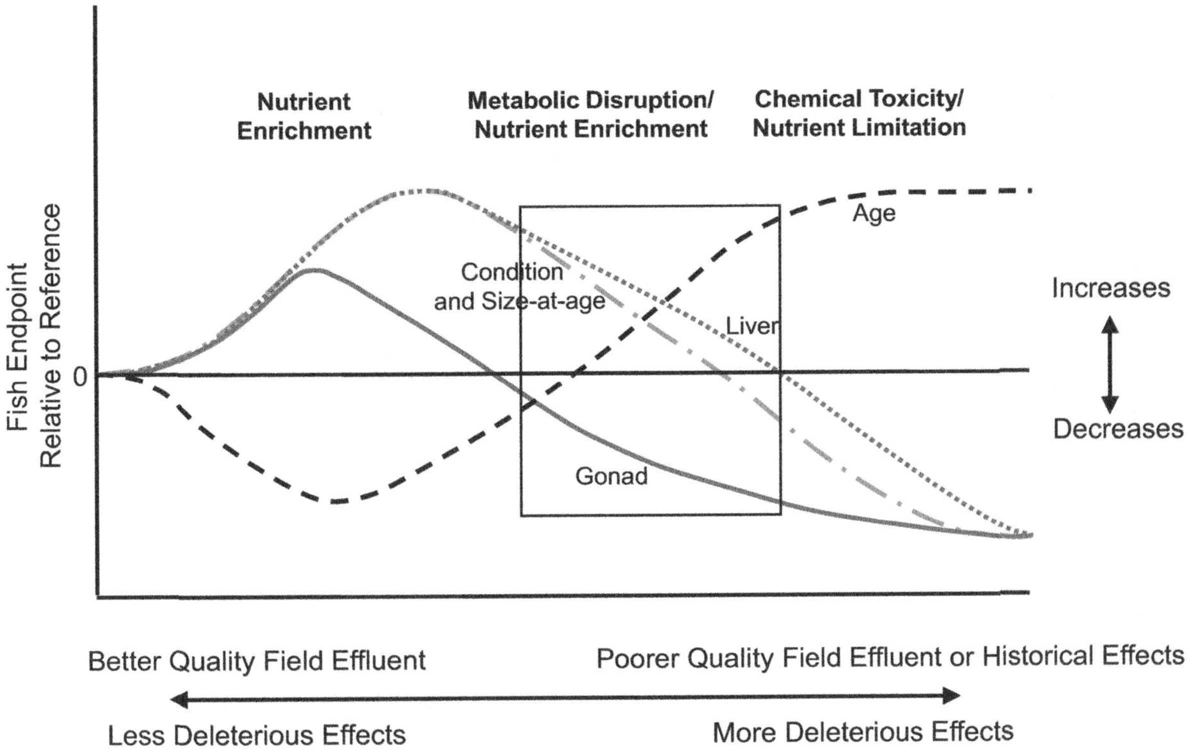

Figure 4-4. Potential Fish Responses to Changes in Nutrients or Chemical Pollutants/Toxicity, Including TOrCs. The box indicates the combination of gonad, liver, and condition responses where TOrCs could be a cause of the responses observed. (From K. Munkittrick, unpublished)

Table 4-6. Summary of Potential Fish Responses to Different Types of Stressors.

Fish		Eutrophication	Toxicity	Endocrine Disruption	Endocrine Stimulation
Survival	Species	Decrease	Decrease		
	Adults	Increase	Decrease		
	Young	Increase	Decrease	Decrease	
Energy use	Altered growth	Increase	Decrease	??	Decrease
	Altered gonad size	Increase	Decrease	Decrease	Increase or no change
	Altered egg production	Increase	Decrease	Decrease	Increase or no change
Energy storage	Altered liver size	Increase	Decrease	Increase	Decrease
	Altered condition	Increase	Decrease	Increase	Decrease or no change
Abnormalities	Intersex			Positive or no change	
	Biomarkers			Positive	Positive
	Lesions, tumours		Increase		

CHAPTER 5.0

WERF TORC DATABASE

One of the objectives of this Phase 1 research was to develop a relational database and user interface with which the water resource community could enter, store, and search TOrC exposure and effects data in the U.S. In addition, the database could also serve as a repository of fate and predicted effects information based on the types of information used to prioritize TOrCs in this research (see Chapter 2.0). The following describes the database and user interface developed for this project, information and data layers currently available in the database, how to access and use the database, and potential usefulness of the database to the water resource community.

5.1 Database and User Interface Description

The WERF TOrCs database is designed to be installed and run from a server connected to a network which provides a centralized database with user access using an internet connection and a web browser. The user interface runs inside an internet browser (e.g., MS Internet Explorer); therefore an end user does not need to install any software on their computer (see Section 5.3). The complete application consists of a Microsoft SQL Server database and a web based user interface developed with Micro Soft .Net C#. The back-end database is relational with linkages and keys between tables that enforce referential integrity so that duplicate data entry and incomplete records are reduced or eliminated. Tables, columns, keys, and descriptions are listed in the data dictionary Appendix (B). The user interface includes a number of tools that allow the user to:

- Upload and download data from a spreadsheet
- Browse for data using a map based interface
- Search (query) for data by a number of parameters including compound, taxonomy, and physical habitat parameters
- View data in tabular format and drill down through the data starting with monitoring location, then samples, and then results
- Manage data, including editing existing data and adding new data
- Manage look-up, or meta-data lists, such as compounds of concern, and analysis methods.

5.2 Information Currently Stored in the Database

The WERF TOrC database contains records of TOrC data measured by several different organizations in surface waters of the U.S. Currently, TOrC data are included from 11 published studies representing a total of 368 unique locations (see Figure 5-1). For each data record, this information includes the TOrC name, CAS number, concentration and concentration units, location, date, and link to the data source citation. In addition, approximate sampling locations have been included for an additional 16 published studies in which TOrC data were collected and measured but data were not reported by site. In these cases, actual data by site are unavailable and therefore cannot be displayed, but source information is made available to the user.

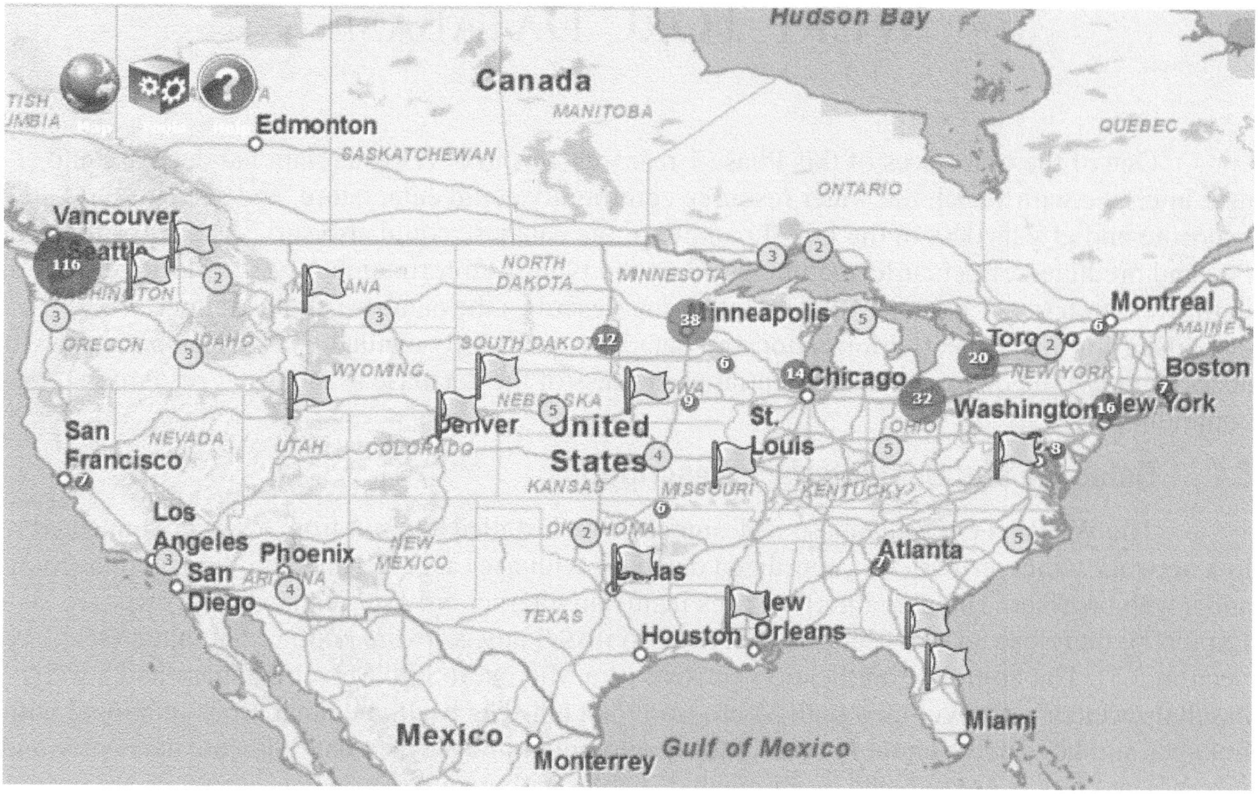

Figure 5-1. Map Showing Current Locations of TOrC Data in the WERF TOrC Database.

In addition to measured TOrC data, the database includes useful fate information, obtained from EPA's most recent version of EpiSuite (e.g., log K_{ow}, half-life in water, K_{oc}) for each TOrC. This is the same fate information used to prioritize TOrCs as discussed in Chapter 2.0. Predicted chronic toxicity information is also included for each TOrC based on EPA's ECOSAR database, where such information is available. Estrogenic activity level for each TOrC is available as well if available. Thus, a user can obtain predicted fate and effects information for many individual TOrCs reported for a given location and time.

Aside from TOrC concentration, and fate and effects information, the database has the ability to store and display biological data (at all biological levels) and habitat data if available. Currently biological data are included in the database for many sites in King County, WA only. By having the capacity to link TOrC, habitat, and biological data for various locations, the database has the potential to serve as a useful analysis tool to help users examine relationships between TOrCs and biological measures. Other data layers currently available in the database include the National Hydrology Data layer, which provides a map of surface water drainages, transportation roadways, and locations of towns and cities.

5.3 Accessing the Database

The database can be accessed with an internet connection and a web browser. It can currently be found at http://werf2.tetratech-ffx.com/. The database is currently stored on a Tetra Tech, Inc. server and will be moved permanently to a WERF server in the near future.

5.4 Searching the Database

After accessing the database, it can be searched by clicking on the "Search Data" option in the green menu bar. The user will see a map (Figure 5-1) and be able to zoom in and out. The map displays interactive flags and circles (Figure 5-1). A flag denotes a sampling location while a circle denotes several sampling locations too close in proximity to be displayed individually. When the user zooms in on a circle location, flags will appear when the user has zoomed in close enough. When the user puts their cursor on a flag, the literature citation, location, and chemical, biological, and habitat data will appear for that location. The map can be viewed as a street map, satellite map, or a USGS map and there are optional map layers that can be applied. The "Search Data" function also allows the user to filter sampling locations by addresses or map selections, by selecting compounds, and by general attributes, such as sampling date, whether the study has been published, or if the compounds sampled include an endocrine disrupting compound. There is a "Tools" icon which can be used to create a summary report. This feature creates an Excel spreadsheet including all types of the information in the database (e.g., citations, chemical information, locations, and chemical results). Once the user has filtered the data the summary report will include only the desired information. If the user has not filtered the data, the summary report will include all information in the database.

5.5 Adding Information to the Database

To add information to the database the user must first register. The user will be required to log-in each time they desire to add or edit data. Once logged in to the database, data input begins by choosing the "Manage Data" option in the green menu bar. There are several options for adding data. The first option is to import the data. This feature allows users to upload data in spreadsheet form to be imported into the database. The second option is to browse the data in the system and manually drill down through the menus to add a study, followed by monitoring locations, sampling events, and finally the actual chemical, biological, or habitat data. It should be noted that data cannot be entered without a location.

5.6 Potential Uses of the WERF Database

Using the search function of the database there are several potential uses. One potential use is the ability to match TOrC data with biological and habitat data. By using the "select compounds feature," the user can filter the results based on select parameters and create a summary report for analysis. For example, filtering for the TOrC acetaminophen, the database reduces the number of flags on the map and produces a summary report with 766 sampling events. Another potential use is being able to view what TOrCs are monitored together. The user can look at multiple TOrCs using the filtering feature or simply put their cursor on the flags to investigate individual sampling locations. Furthermore, the database could be used to investigate where TOrCs of interest are monitored and if these TOrCs are more likely to be found in one geographical location over another. The database could also potentially be used to see trends in TOrC monitoring or occurrence. If the database were widely used and contained large amounts of monitoring data over many years, it would be possible to analyze the data for a general or site specific increase or decrease in TOrCs of interest. The more data in the database, the more potential uses it will have.

CHAPTER 6.0

WHERE DO WE GO FROM HERE?

6.1 Critical Data Gaps and Uncertainties

While much has been learned in this research, the foregoing chapters of this report highlight several critical information gaps that currently limit accurate assessments of TOrC risks in aquatic systems, and specifically, the extent to which TOrCs are or could be a critical stressor. Knowledge gaps identified in this research affect our ability to: 1) confidently identify those TOrCs that should be given the greatest attention in terms of aquatic ecological risks; 2) accurately screen or rank sites in terms of their potential TOrC risks to aquatic life; and 3) diagnose whether certain TOrCs, or types of TOrCs, caused or could cause impacts on aquatic populations or communities. The following briefly summarizes the critical knowledge gaps that have been identified pertaining to these three broad areas.

(1) Which TOrCs Should be Given the Greatest Attention?

The prioritization approaches developed in this research provide a useful starting point in terms of identifying TOrCs that have the potential to be present in aquatic environments at concentrations that could cause biological effects. However, there are many uncertainties regarding the available information with which to prioritize TOrCs. For example, maximum concentrations of certain TOrCs reported in the literature may be inaccurate; i.e., chemical analyses may have either over-estimated or under-estimated TOrC concentrations in some monitoring studies resulting in both false positives (TOrC thought to exceed toxicity or endocrine activity thresholds when in fact it did not) and false negatives (TOrC thought to never exceed thresholds when in fact it has). As discussed in Chapter 2.0, uncertainties in low level analytical procedures for certain TOrCs that are currently being addressed by many organizations. In addition, there are significant gaps regarding toxicity and estrogenic activity data for several TOrCs identified in our occurrence database. As a result, some TOrCs were not capable of being considered in any of the three prioritization approaches. This issue is even further accentuated in terms of gaps regarding general endocrine effects of TOrCs apart from estrogenic activity. Information on thyroid disruption effects in fish, for example, is known for a handful of the TOrCs identified in our occurrence database and, as yet, there is no readily available "activity equivalents" process that can be used similar to that used to rank estrogenic activity of TOrCs.

Relating modes of action to ecotoxicology assessments is an area of active research. For example, the Adverse Outcome Pathways approach (Ankley et al., 2009) uses a mode of action approach, which attempts to assess chemicals based on the biochemical pathways affected, and uses tissue and cellular techniques to help diagnose effects on an organism. This approach, as well as the intelligent testing strategy (Bradbury et al., 2004; ECETOC, 2007) and genomics (Ankley et al., 2006; Poynton and Vulpe, 2009) promise to help fill some of the information gaps regarding potential effects of various TOrCs and may provide ways to perhaps refine the prioritization process and diagnostic capabilities.

(2) What Types of Sites Are Most at Risk Due to TOrCs?

This research produced a preliminary screening approach for wastewater influenced sites that takes into account three types of factors: (a) source of influent to the WWTP; (b) type of treatment system used and quality of the effluent; and (c) ecological characteristics of the site that may affect organisms and their exposure to TOrCs discharged from a point source. These three factors, along with available biological and chemical data from a site, were applied as a screening tool to several case studies as discussed in Chapter 4.0. However, the measures included thus far for the influent and WWTP factors in the screening tool, are based largely on personal experience of researchers and limited published information. One could view each of the measures in the screening tool as a hypothesis needing to be tested. While results of applying the screening tool appeared to be consistent with more in-depth findings for the seven case studies examined, in fact only in the Boulder WWTP and the Kitchener WWTP case studies did in-depth analyses indicate that certain TOrCs were in fact a potential cause of biological impairment at the site.

The screening tool, as developed thus far, appears to be applicable to many types of potential wastewater-related stressors such as ammonia, high BOD, legacy contaminants, and metals, rather than TOrCs per se. Refinements of the screening tool using the iSTREEM model appears to be a fruitful avenue worth pursuing further because the model attempts to quantify, at least in relative terms, various types of potential stressors at a site such as nutrients, habitat impairment, and pesticides, as well as many high priority TOrCs that may be discharged from a WWTP. While the model approach cannot replace on-site data collection and analysis, it may help prioritize sites for further evaluations.

Regardless of whether TOrC exposure is modeled at a site (e.g., using the iSTREEM model) or inferred using various surrogate indicators (e.g., via influent and WWTP characteristics), certain assumptions are used regarding relationships between the type of treatment and TOrC exposure in the waterbody. The companion WERF project (CEC4R08) dealing with treatability of TOrCs has determined that relationships between influent characteristics, WWTP characteristics, and effluent TOrC concentrations are extremely complex and still poorly understood. This, and other recent research (e.g., Teske and Arnold, 2008; Guo et al., 2009; Phillips et al., 2010; Dunlavey et al., 2010) should shed light on influent sources and treatment processes that are more likely to be associated with high TOrC risks to aquatic life.

(3) Do Certain TOrCs Present an Unacceptable Risk to Aquatic Populations and Communities at My Site?

Eco-epidemiological approaches developed thus far such as EPA's CADDIS, and Canada's EEM, appear well-suited to evaluate common types of stressors such as excess nutrients, habitat quality impairment (such as sedimentation or riparian corridor degradation), and pollutants that exhibit classical toxicity responses (e.g., copper). However, our evaluation of several case studies in this research indicates that far more site information, and different types of data, are required to either eliminate TOrCs as a candidate cause of observed impairment or to demonstrate that certain TOrCs are a likely cause of impairment.

Both CADDIS and EEM are flexible in their approach to accommodate TOrCs but the type of site data typically available is insufficient for evaluating TOrCs as a candidate cause. For the types of more common stressors mentioned above, effects of these stressors are often captured using standard bioassessment techniques. Species sensitivity information is often available with which to evaluate these stressors. For most high priority TOrCs identified in this research, standard bioassessment metrics and indices do not provide definitive information that is diagnostic of TOrC effects; most metrics and indices respond to many stressors and are therefore not truly diagnostic. Other types of analyses are required to diagnose organism exposure to certain types of TOrCs (e.g., elevated vitellogenin levels in male fish are diagnostic of some estrogenic TOrC(s) at the site).

A critical information gap arises from the fact that diagnosis of TOrC exposure is generally accomplished by examining effects at the individual organism level, yet we are interested in diagnosing TOrC effects at the population or community level. There are many potential steps between organism exposure/effects and population level effects (see Figure 6-1), at least some of which will be dependent on site factors such as those included in our screening approach. As we noted in Chapter 4.0, developing a site-conceptual model with potential stressor sources, candidate causes, causal pathways, and ecological effects (both indirect and direct effects) is a useful tool to help document hypotheses, highlight critical data gaps in diagnosing causes of known impairment, and identify potential effects of TOrCs. However, given the current state of knowledge, there may be many alternate pathways, most of which have little actual evidence thus far.

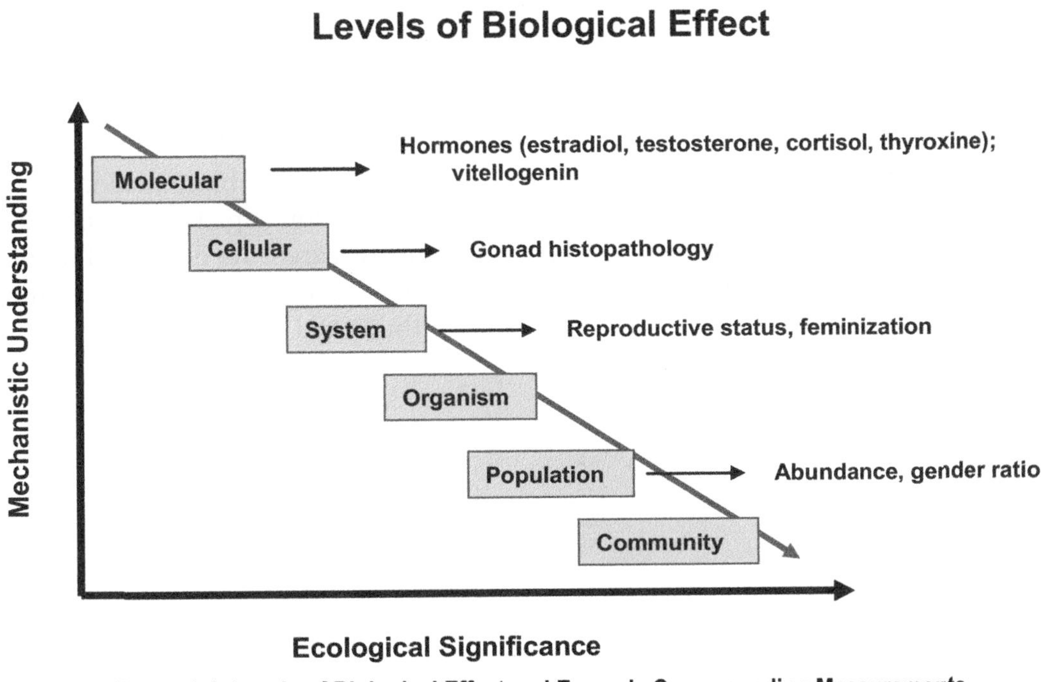

Figure 6-1. Levels of Biological Effect and Example Corresponding Measurements.
Each level has different characteristics of sensitivity, interpretation, and ecological relevance.
From Bay 2010 and used in Diamond et al., 2010d.

Figure 6-2 presents a conceptual model for a highly estrogenic TOrC such as EE2 illustrating some of the pathways that are likely to occur if exposure is sufficient. Note that exposure and effects information is required at multiple biological levels to test this conceptual model. The experimental lake study by Kidd et al. (2007) generally corroborated many of these factors using several doses of 5 ng/L EE2. In most TOrC-related monitoring studies conducted thus far in the U.S. and elsewhere, such multi-level information is typically lacking.

Figure 6-3 adds an environmental context to the conceptual model in Figure 6-2, highlighting several factors that could affect the extent of exposure and therefore, the severity of the effects identified in Figure 6-2. Factors include the extent of immigration and emigration of organisms among different areas, the abundance of refugia from TOrC exposure, and the existence of other stressors in the system. Some of these factors are included in the screening approach, as discussed in Chapter 3.0 but the extent to which these factors can mitigate potential effects of estrogenic TOrCs is unknown at present.

The above conceptual models illustrate several points. First, exposure information without population-level effects information is of limited use in furthering our understanding of TOrC effects, types of sites or populations most at risk due to TOrCs, as well as levels of TOrCs that may or may not pose a risk. Second, physiological and histological indicators may provide more useful diagnostic exposure information than measurements of a partial list of TOrCs monitored infrequently. Third, there are currently many uncertainties regarding the effects of mixtures of TOrCs and other chemicals on the pathways shown, as well as the degree to which perhaps other pathways become more critical. Making the linkages between TOrC exposure and effects at multiple levels of biological organization is perhaps the biggest challenge to accurately assess ecological risks of TOrCs.

6.2 Testable Hypotheses for Phase 2 Research

The critical uncertainties and information gaps summarized in the previous section were organized by three stages within the overall process of assessing risks of TOrCs on aquatic populations and communities. The critical uncertainties from each of those stages have been organized around three themes which we believe are key issues of concern to utilities and the water resource community in general: 1) site screening; 2) determining where TOrCs present an unacceptable risk; and 3) diagnosing whether TOrCs are responsible for biological impact. The researchers have translated what they believe are the key information gaps for each of these themes into hypotheses that could be tested in Phase 2 of this research. As discussed in the next section, it is important that Phase 2 research is staged such that practical solutions regarding some basic aspects of TOrC uncertainty are available to utilities and others almost immediately after the start of Phase 2.

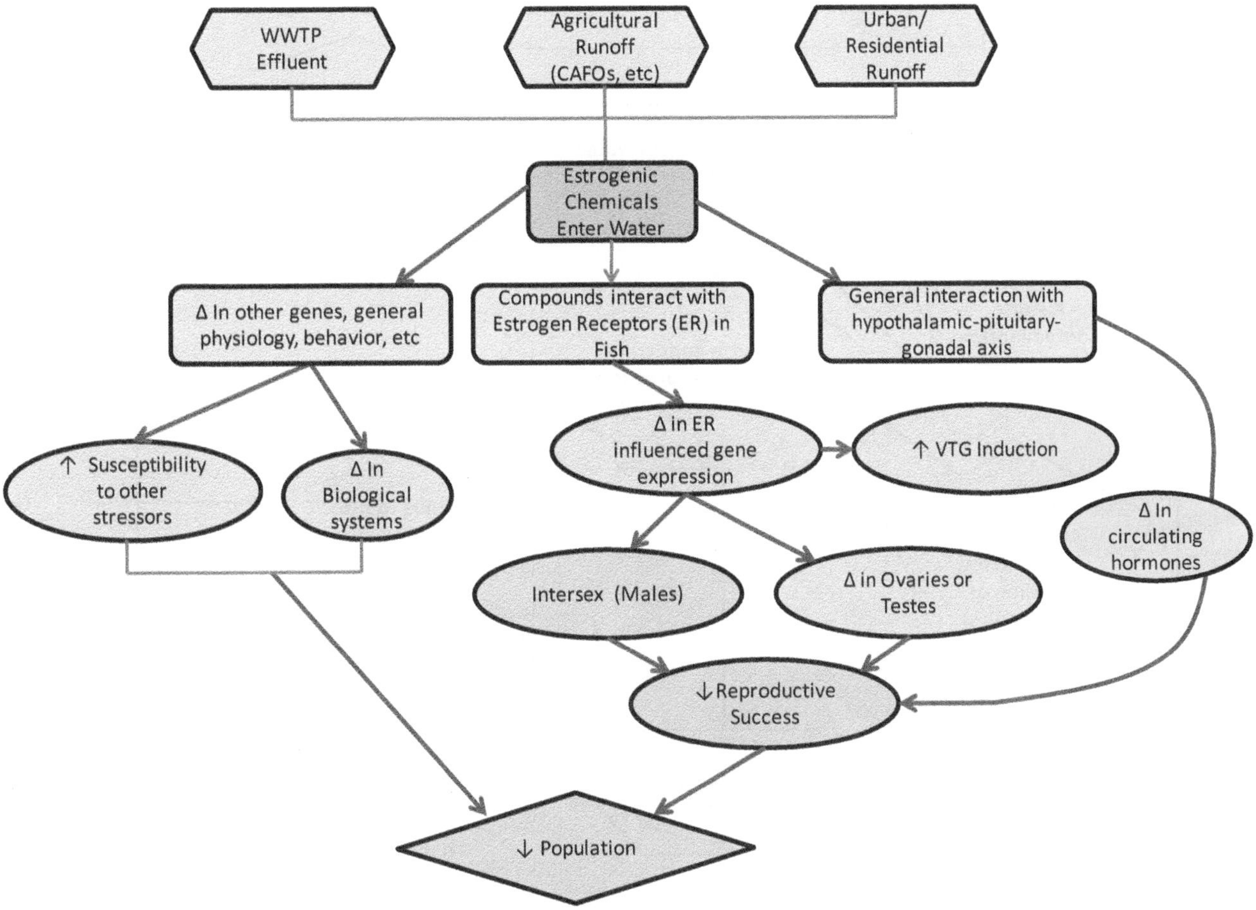

Figure 6-2. Conceptual Model Showing Potential Pathways Involving an Estrogenic TOrC Such as 17-α Ethynylestradiol (EE2).

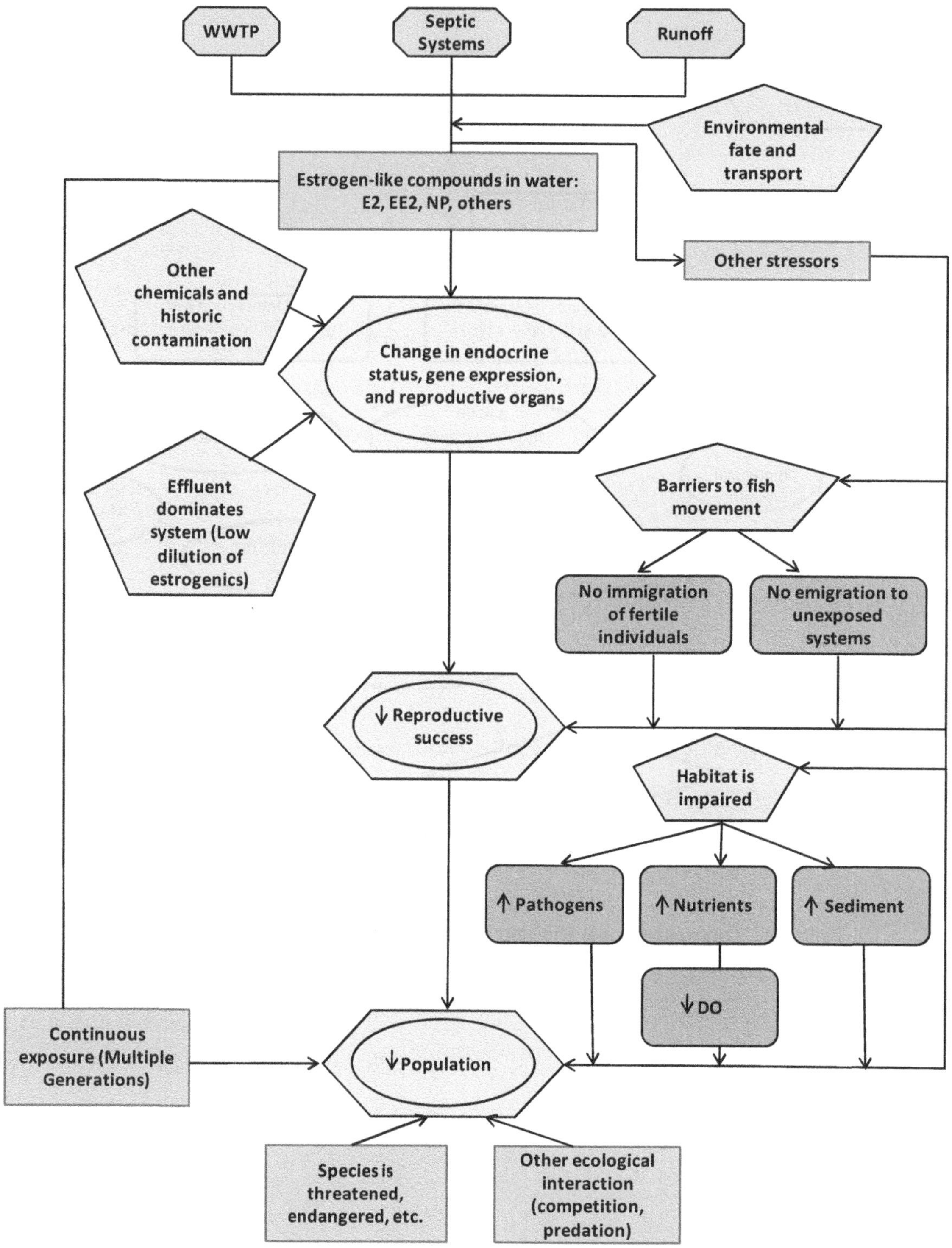

Figure 6-3. Conceptual Model Showing Additional Factors that Could Influence the Extent or Presence of Effects on Fish Due to Exposure to an Estrogenic Chemical Such as 17-α Ethynylestradiol (EE2).

Theme 1 – Site Screening/Site Ranking Process

Overarching question: *Under what conditions do waterbody factors or WWTP factors drive aquatic ecological risks of TOrCs?*

General approach: Evaluate a series of sites with different potential levels of risk from TOrCs to distinguish high from low risk sites based on WWTP and receiving stream characteristics.

Probability of success: High
Complexity: low
Cost: low

This theme addresses questions of relevance to utilities and may quickly provide a more refined tool that could be applied to screen or rank sites in terms of relative risk due to TOrCs. This tool could also provide guidance regarding potential wastewater engineering solutions that might tend to lower TOrC risks to aquatic biota. This theme does not deal with the specific issue of cause or how TOrCs may be contributing to impairment. These issues are addressed in subsequent themes.

Specific testable hypotheses in Theme 1:

Ho 1: Smaller streams and/or WWTP discharges with little dilution under base flow conditions are more likely to exhibit ecological effects due to TOrCs than larger streams and rivers, or WWTPs with much greater dilution.

Effects of TOrCs (as with all chemical pollutants) are dependent on exposure duration and frequency as well as the magnitude of concentrations occurring in the waterbody (Diamond et al., 2006). However, unlike most chemicals currently regulated in the U.S., TOrCs that are EDCs may require longer or more frequent exposures than typically considered in order to elicit population-level effects. This hypothesis tests the degree to which one of the waterbody factors in the current screening tool, effluent dilution instream, is an accurate site screening indicator. Effluents having greater dilution (and perhaps greater initial mixing) are hypothesized to be less likely to have TOrC risks because TOrC concentrations will be lower, with lower organism exposure to TOrCs. Having predicted, or preferably measured effluent and stream concentrations of high priority TOrCs (e.g., the 41 TOrCs identified using prioritization Approach #1 or the 23 TOrCs identified using all three prioritization approaches) will enhance the test of this hypothesis, as well as the hypotheses in Theme 2 below. This is because dilution by itself may not explain TOrC risk if effluent or upstream concentrations of certain TOrCs are high enough such that, even with available dilution, TOrC concentrations exceed potential risk levels.

A corollary of this hypothesis is that larger, open aquatic systems, such as large rivers and estuaries, are less likely to exhibit effects of WWTP-derived TOrCs than smaller streams and closed waterbodies such as ponds and small lakes. The reasoning, similar to that discussed above, is that larger waterbodies have more opportunity for dilution (and perhaps degradation) of TOrCs than smaller waterbodies and open waterbodies are likely to provide greater opportunity for organism refugia or organism avoidance of TOrCs than closed systems, thereby limiting organism exposure to TOrCs. This latter assumption is more directly tested in the next hypothesis.

Ho 2: Receiving waterbodies having barriers to fish movement, low flow, few refugia, or other factors favoring exposure of biota to high priority TOrCs (e.g., continuous WWTP discharges vs intermittent discharges) are more likely to exhibit risks from TOrCs.

As noted in the discussion under the first hypothesis, any factors that are likely to reduce exposure of organisms to TOrCs, should result in less TOrC risk to biota. Thus, sites having natural or human-made characteristics that tend to maintain greater exposure of biota to WWTP-derived TOrCs (or TOrCs originating from other sources) are hypothesized to be at greater risk due to TOrCs. The current spreadsheet tool incorporated this factor as part of the site ecological factor area. While a couple of the case studies yielded some evidence supporting this factor as a useful indicator of TOrC risk (Kitchener, Ontario and Boulder Creek, CO), there were too few sites to explore this hypothesis in depth.

A related hypothesis is that smaller wastewater plants, having generally lower removal efficiency for compounds such as TOrCs, may be at greater risk of TOrC effects than much larger WWTP plants that tend to have greater hydraulic retention time and more treatment steps that tend to degrade or remove TOrCs prior to discharge. All of the case studies examined had continuous discharges of treated effluent. Therefore, the effects of intermittent discharges (e.g., smaller package wastewater treatment units) could not be evaluated.

Theme 2 – Where Do TOrCs Present an Unacceptable Risk to Aquatic Biota?

Overarching question: *In what types of sites are "high" TOrC concentrations associated with observed biological effects?*

General approach: Evaluate impacts at a set of sites with similar physical characteristics and varying levels of high priority TOrCs, including those with concentrations far exceeding conservative thresholds

Probability of success – high
Complexity: medium
Cost: low

This theme will provide correlative evidence of concerns related to TOrCs, focusing efforts at sites with the highest potential exposures to high priority TOrCs. This theme may address the question "how much of a biological problem are sites with elevated TOrC concentrations", assuming other stressors are equivocal across sites. This theme would examine high priority TOrCs identified in this research as noted under Theme 1. Results of exploring this theme will be correlative and interpretation may be confounded by other water quality issues. Also, the experimental design for this theme may require pre-existing knowledge about facilities and their effectiveness at removing high priority TOrCs. Some of this knowledge may be available through information gained in the companion WERF project (CEC4R08).

Specific testable hypotheses of interest:

Ho 3: Sites with predicted or observed total equivalence (as TUs or Eeqs) of high priority TOrCs >> 1.0 are more associated with reproductive, growth, or other population-level effects as compared to sites with equivalents < 1.0

If certain TOrCs are indeed causing impairment of biological populations and communities at a site, there should be some correlation between cumulative TOrC risk, expressed in terms of total estrogenic equivalents (EEq, based on EE2 or E2) and/or total toxicity units TUs based on ECOSAR predicted chronic toxicity levels for algae, invertebrates, and fish), and biological effects. In this hypothesis, total TUs or EEqs at a site are based on either predicted effluent values for high priority TOrCs (using the iSTREEM model for example) or preferably, measured effluent and stream values. A range of sites having different levels of biological impairment, and preferably, different levels of high priority TOrC concentrations would be needed to address this hypothesis. Ideally, there would be few other obvious stressors at many of these sites in order to reduce confounding factors such as excess nutrients or habitat quality impairment. This hypothesis could not be evaluated using the case studies examined in this research because TOrC concentration data were available for only a few sites, two of which were based on predicted concentrations using the iSTREEM model. POTW TOrC data being generated by Oregon DEQ in 2010 could provide some useful case study sites in this regard based on preliminary discussions between Tetra Tech, Oregon Association of Clean Water Agencies (ACWA), and Oregon DEQ staff.

Ho 4: Sites with high concentrations of TOrCs that have direct chronic toxicity effects result in greater population-level effects as compared to EDC TOrCs, based on concentrations expressed as total equivalents (either TUs or EEqs)

The prioritization process used by Tetra Tech revealed that the most sensitive ECOSAR-derived chronic toxicity endpoint was a more sensitive indicator of risk potential for most TOrCs than estrogenic activity, given the maximum concentrations of TOrCs identified in our occurrence database. In addition, TOrCs whose mode of action is direct toxicity are postulated to have greater effects on populations than EDCs because effects may be more subtle and, in some cases, reversible if given sufficient time without EDC exposure. Using TOrC information collected under the previous hypothesis, high priority TOrCs could be segregated into those whose MOA is primarily endocrine disruption versus those whose MOA is direct toxicity, via various mechanisms such as suppression of neural pathways or immunosuppression. Summing the total equivalents of TOrCs at several sites (as TUs or EEqs), it may be possible to correlate specific types of TOrCs with observed measures of biological condition.

Theme 3 - Are TOrCs Responsible for Impacts?

Overarching question: *Can a retrospective, stressor-identification approach partition the effects of TOrCs from other stressors or will the approach be process of elimination where TOrCs are the most likely remaining cause of effects?*

General approach 1: Find sites scheduled for improvements, or sites with historical data that have undergone improvements, and look for improvements in responses that could be related to TOrCs.

General approach 2: Evaluate sites where impacts have been found, and where TOrCs and other stressors are present and try to identify whether TOrCs are a cause of impacts based on multiple biological endpoints.

Probability of Success: medium
Complexity: high
Cost: high

This theme focuses efforts on the process used to diagnose whether impacts are due to TOrCs. Sites having other more serious causes of impact can be identified and prioritized to obtain the most environmental benefit. Investigation of cause, however, can be expensive and time consuming.

Specific testable hypotheses of interest:

Ho 5: Sites having few potential stressors other than TOrCs are more likely to yield a definitive answer as to whether TOrCs are a cause of observed biological effects than a site having multiple stressors other than TOrCs.

The various case studies examined in this research demonstrated the complexities involved in attempting to diagnose whether observed impacts are (or were) due to TOrCs. While there was some variation in terms of number of co-occurring stressors in the case studies examined, available data were too sparse to test this hypothesis. The Boulder Creek case study, one of the few sites thus far available in the U.S. in which there is biological information "before" and "after" WWTP upgrade, examined TOrC concentrations and estrogenic effects on a single fish species but not population or community level effects. There may be other sites with "before" and "after" data that may become available soon and which can be used to test this hypothesis. Sites having multiple stressors in addition to TOrCs, particularly stressors that are known to cause large-scale changes in biota (e.g., eutrophication, sedimentation), may be unlikely to yield definitive statements regarding TOrC risk because those effects are being masked by much more intense disturbances at the site. By examining a range of sites where impacts have been observed (preferably different types of impacts and/or impacts affecting different assemblages), along with different co- stressors, it should be possible to apply various diagnostic tools to determine if TOrCs are responsible for observed impacts. However, as shown in the evaluation of the various case studies, this will be possible only if measures of exposure and effect are examined at multiple biological scales.

Ho 6: Sites with a high incidence of TOrC-induced intersex in a fish population are associated with population-level effects such as reduced abundance, biomass, and skewed age distribution.

The Boulder Creek case study demonstrated linkages between effluent TOrCs, expressed as EEq, estrogenic activity in white sucker individuals, and intersex frequency in that population. However, the final linkage to population/community effects has not been examined thus far. The other case studies examined had even less evidence regarding linkages between known TOrCs, individual effects, and population effects. Biologically, it stands to reason that if a population of fish is composed of almost entirely intersex individuals, normal reproduction is not possible and changes in the population should be observed such as reduced fecundity, reduced abundance, and skewed age distribution. Yet, to date there have been few if any clear examples demonstrating both a high intersex rate and a population-level effect in non-experimental studies (e.g., Kidd et al. 2007). Repeated studies in the Potomac and Shenandoah Rivers in West Virginia, for example, have demonstrated high intersex frequencies in smallmouth bass as well as unexplained periodic fish kills (Blazer et al., 2007; Alvarez et al., 2009) and yet clear population-level effects have not been identified thus far. Part of the issue is that most sites have not been examined in a methodical, multi-disciplinary way that addresses multiple biological scales. By conducting such studies at multiple sites with a documented high incidence of EDC-induced reproductive effects on certain fish species, it should be possible to address this hypothesis.

6.3 Strategy for Phase 2 Evaluation of Critical Hypothesis

Any in-depth study of TOrC risks in aquatic systems will be complex and require various levels of sophistication in terms of data collection and analyses. It is critical that Phase 2 is strategically formulated such that useful guidance or procedures are obtained in the short-term as well as throughout the research effort.

One approach for addressing the nearly intractable complexity of establishing causal relationships among TOrCs (both individually and as mixtures) and effects on aquatic populations and communities is to design and conduct Phase 2 research along the continuum of complexity, cost, and causality (Figure 6-4). The three research themes summarized in the previous section follow a continuum from low to high complexity (screening level analyses to established causal relationships), and low to high costs. The first theme is to test the screening approach developed in Phase 1 to partition wastewater treatment facilities and receiving streams into low, moderate, and high risk sites. Research for the second theme looks more closely at a subset of the screening sites to establish associations across a gradient of TOrC concentrations and biological effects (i.e., from physiological and histological changes in organisms to changes in aquatic population and community indicators). With a better understanding of correlative or associative changes, the third theme is designed to establish causal relationships among TOrCs (individual chemicals and mixtures) and aquatic population and community indicators.

Within each of these themes, we proposed two testable hypotheses that we believe are high priority from a user standpoint although a series of research questions or hypotheses could be identified to focus the research and increase the likelihood of providing tools and procedures useful to wastewater treatment facilities and other WERF subscribers in assessing the potential risk of TOrCs at their sites (Figure 6-5). These products and tools would emerge along the continuum as research is conducted. The research questions or hypotheses within each theme could be listed in hierarchical order so that the answer to each question builds on previous information to increase both the effectiveness and efficiency of the research (See Section 6.2).

Study designs for Phase 2 research could follow the research themes and the continuum of complexity, cost, and causality (Figure 6-4). Theme 1 research would involve screening a wide range of wastewater treatment facilities and receiving streams from those expected to be at low risk on one end of the continuum to those expected to be at high risk at the other end of the continuum (Figure 6-5). This would permit using a consistent, comparable, and compatible approach for characterizing both the wastewater treatment facilities and the receiving water bodies not only with respect to TOrCs, but also other stressors and environmental factors, as well as biological effects at the organism, population and community level. The study design for Theme 2 research might be to select a subset of the screening sites to look for associations between TOrCs and biological effects indicators, minimizing the confounding effects of other stressors and environmental factors (Figure 6-5). The study design for Theme 3 research would be to select a few sites from both the high risk and low risk site categories and conduct intensive research to establish causal relationships among TOrCs and biological effects at the organism, population, and community levels modified by other stressors and environmental factors (Figure 6-5). Associative relationships established during the Theme 2 research would be used to guide causal research at the high risk sites and test these same relationships at low risk sites. The Theme 3 research could be extended to moderate risk sites if resources and funding was available.

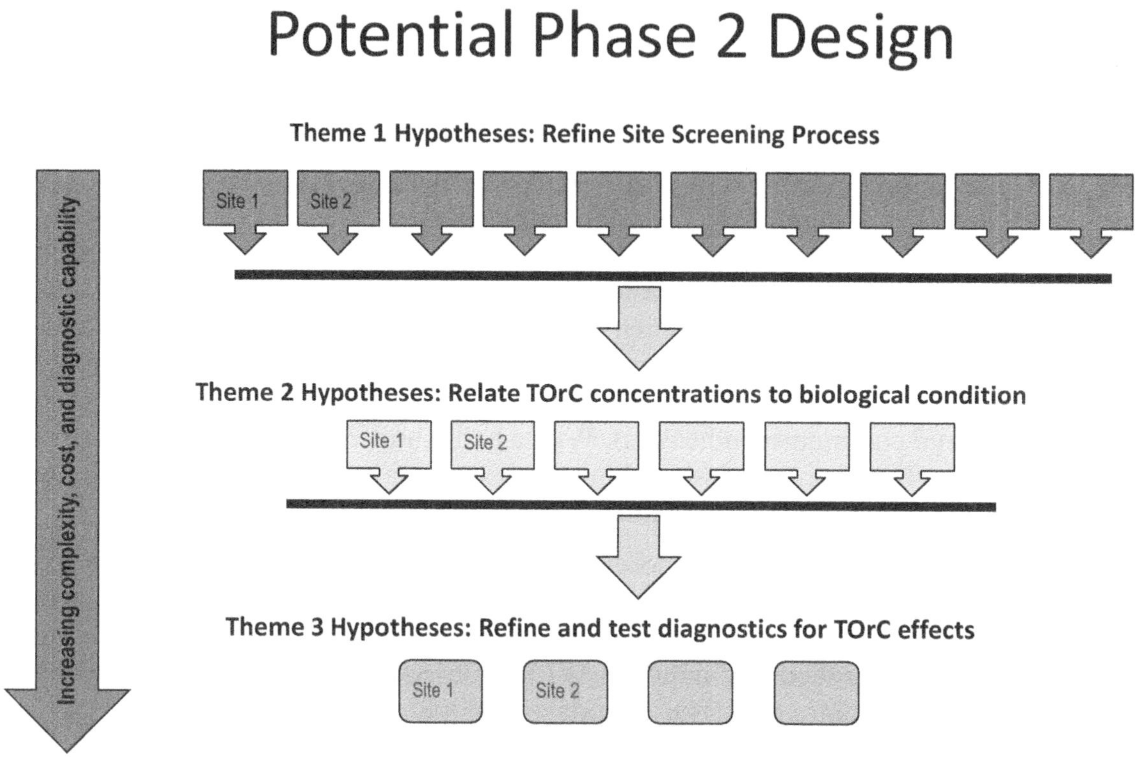

Figure 6-4. Proposed Phase 2 Continuum of Research Projects and Products at Different Levels of Complexity, Causality, and Cost.

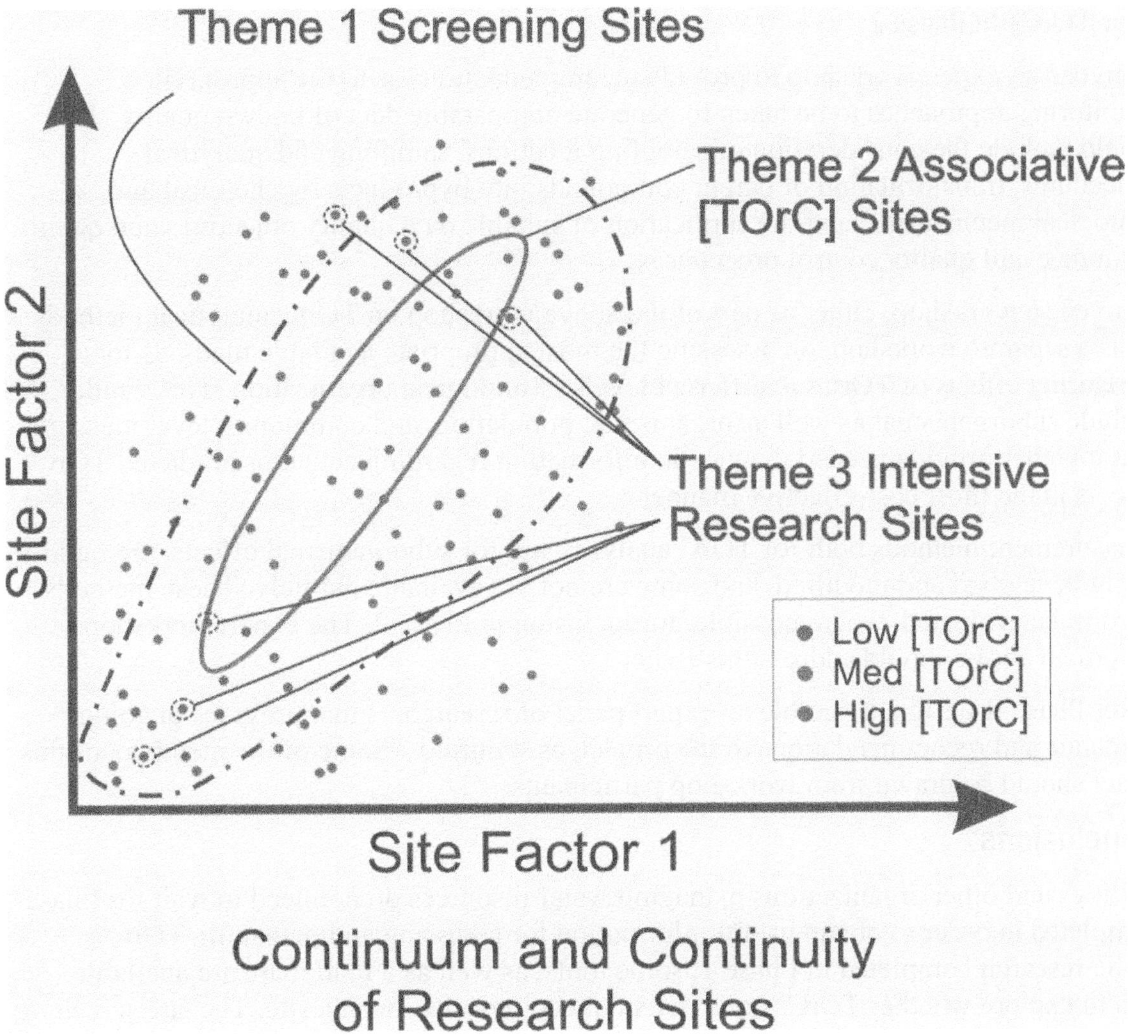

Figure 6-5. Nested Approach to Phase 2 Project Sites that Builds on Information Gained Through Screening, Associative Analyses, and Intensive, Causal Research at These Sites.

The above approach would permit research partners and collaborators to participate anywhere along the continuum or within any of the three research themes. This would also permit WERF to identify research not being addressed by its partners and collaborators and target these areas for funding.

As discussed in Chapter 2.0, sampling and analysis methods for certain types of high priority TOrCs are still being actively researched or evaluated in multi-laboratory testing. It is critical that TOrC data obtained in Phase 2 of this research are accurate and representative. To this end, there are several recommendations specifically regarding sampling and analytical methods for TOrCs in Phase 2:

1. Convene an expert workshop to provide recommendations as to the appropriate monitoring approaches to be taken to generate comparable data of known quality. This would include the consideration of sampling locations, sampling and analytical procedures, transformation of parent compounds into byproducts by chemical and biological mechanisms, and the application of suitable data quality objectives and quality assurance and quality control procedures.
2. Convene a workshop, either as part of the above workshop on TOrC analytical methods or as a separate workshop, on assessing the most appropriate available methods for measuring effects of TOrCs at different levels of biological organization. This would include suborganismal as well as organismal, population, and community level measures that together provide needed diagnostic information regarding actual or predicted TOrC effects in the most cost-effective manner.
3. Measurement methods both for TOrC analyses and for suborganismal effects, are rapidly being developed and modified, and some are not yet available publicly. These methods need to be evaluated, where possible, for inclusion in Phase 2. The expert workshops described above should address this issue.
4. After Phase II design, assemble an expert panel of researchers that provides ongoing guidance and recommendations to the project, as it unfolds. Some of the members of this panel should be drawn from workshop participants.

6.4 Conclusions

Utilities and other organizations managing water resources do not need to wait for Phase 2 to be completed in order to obtain useful information for assessing and managing TOrCs. Based on the research completed in Phase 1, some tools, as well as a road map, are available with which to explore whether TOrCs are a potential ecological issue at a site. The site screening tool developed in this research, while simple, may be a useful first step to decide whether TOrCs are a potential cause for concern at a site. The three-step process shown in Figure 6-6 summarizes the general steps that will be useful in assessing TOrC risks at any site. The prioritization approaches developed in this research may be a useful starting point for deciding which TOrCs should be monitored in the receiving waterbody or effluent. Figure 6-7 summarizes some of the general steps that may be useful to help diagnose whether TOrCs are indeed a cause of concern at a site. Some of the analyses suggested are not necessarily costly or time-consuming, depending on the size of the site and the level of expertise available. For example, trained fishery biologists can help assess the degree to which intersex occurs in a given fish population. As more diagnostic biological tools become readily available (e.g., reliable analytical methods for some high priority TOrCs; vitellogenin assays that are cheaper and easier to run; microarrays for genotoxicology assessments of exposure), many of the other analyses suggested should be more accessible to utilities and water resource managers.

Partitioning the ecological effects of trace level organic compounds on aquatic populations and communities will be a long-term process. However, the Phase 1 research has laid a solid foundation for the Phase 2 research. Building on this foundation through Phase 2 research, as described in this chapter, will permit continuous feedback between incremental increases in our understanding of TOrC effects and development of interim tools and the additional risk hypotheses that emerge from the application and testing of these tools by WERF utilities, research partners, and collaborators. The proposed Phase 2 research design should permit all potential research partners to leverage their funds and resources at multiple points along the proposed research continuum. It is through collaborative research that progress will be achieved.

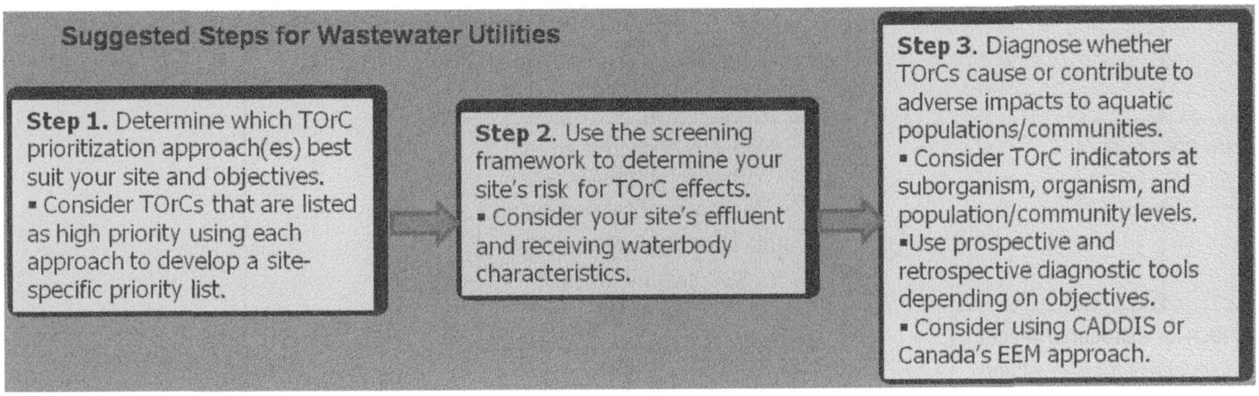

Figure 6-6. General Steps Wastewater Utilities Might Follow to Assess Whether TOrCs Are a Concern at Their Site.

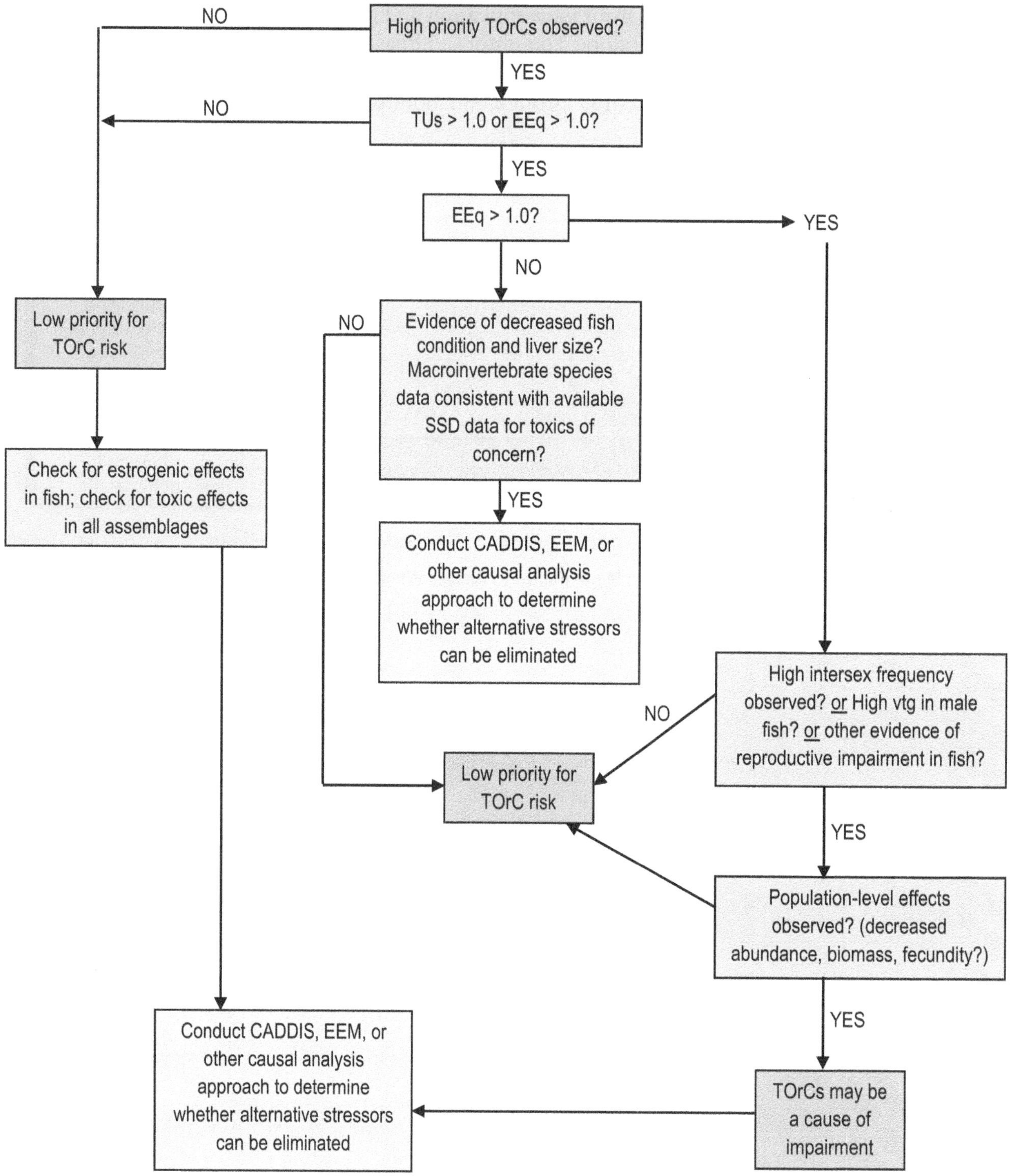

Figure 6-7. Flowchart Summarizing the Key Steps in Determining Whether Trace Organic Compounds (TOrCs) Are or Could Be a Concern at a Site. TU = toxicity units; EEq = estrogen equivalents; SSD = species sensitivity distribution; CADDIS = EPA's Causal Analysis Diagnosis Decision Information System; EEM = Canada's Environmental Effects Monitoring approach; vtg = vitellogenin

APPENDIX A

SUPPORTING INFORMATION FOR HIGH-PRIORITY TORCS IDENTIFIED IN THIS RESEARCH

Table A-1. List of Trace Organic Compounds Identified as High Priority Using at Least One of the Three Prioritization Approaches, Including the Type of Compound, Type of Taxon (algae, daphnid or water flea, or fish) Most Sensitive to the Compound According to EPA's Ecological Structural Activity Database (ECOSAR), and General Mode of Action or Class of Compound if Known.

Prioritization approach 1 = risk based, 2 = risk + persistence and bioaccumulation scores, 3 = persistence, bioaccumulation, and toxicity scores.

Compound	CAS Number	Trace Organic Type	Approach(es) Considered High Priority	Most Sensitive Species (ECOSAR)	Known EDC	Mode of Action or Class of Compound
Acetyl cedrene	32388-55-9	Deodorizer/Fragrance	2, 3	Fish		
Acetyl-hexamethyl-tetrahydro-napthalene (AHTN)	1506-02-1	Deodorizer/Fragrance	3	Daphnid	Yes	
Celestolide (ADBI)	13171-00-1	Deodorizer/Fragrance	2, 3	Fish		
Galaxolide	1222-05-5	Deodorizer/Fragrance	1, 2, 3	Fish		
Musk ketone	81-14-1	Deodorizer/Fragrance	1, 2, 3	Fish	Yes	
Musk xylene	81-15-2	Deodorizer/Fragrance	2, 3	Fish		
OTNE	54464-57-2	Deodorizer/Fragrance	2, 3	Fish		
Phantolide (AHMI)	15323-35-0	Deodorizer/Fragrance	3	Fish		
Tonalide	21145-77-7	Deodorizer/Fragrance	1, 2, 3	Fish		
Traseolide (ATII)	68140-48-7	Deodorizer/Fragrance	2, 3	Fish		
1,1'- 1,2-ethanediylbis(oxy) bis 2,4,6-tribromo-Benzene	37853-59-1	Flame Retardant	3	Fish		
1,4:7,10-Dimethanodibenzo a,e cyclooctene, 1,2,3,4,7,8,9,10,13,13,14,14-dodecachloro-1,4,4a,5,6,6a,7,10,10a,11,12,12a-	13560-89-9	Flame Retardant	3	Fish		
1H-Isoindole-1,3(2H)-dione, 2,2'-(1,2-ethanediyl)bis 4,5,6,7-tetrabromo-	32588-76-4	Flame Retardant	3	Fish		
Hexabromocyclododecane	3194-55-6	Flame Retardant	1, 2, 3	Daphnid	Yes	Endocrine disruptor (thyroid)
PBDE-209	1163-19-5	Flame Retardant	1, 2, 3	Fish	Yes	Endocrine disruptor (thyroid)
1,1'-Biphenyl, bis(1-methylethyl)-	69009-90-1	Industrial Chemical	3	Fish		
1,1'-ethylidenebis 3,4-dimethyl-Benzene	1742-14-9	Industrial Chemical	3	Fish		
1,3-DiNO2 2-CL 5-CF3 benzene	393-75-9	Industrial Chemical	3	Daphnid		

Table A-1. Continued.

Compound	CAS Number	Trace Organic Type	Approach(es) Considered High Priority	Most Sensitive Species (ECOSAR)	Known EDC	Mode of Action or Class of Compound
2-(2H-benzotriazol-2-yl)-4,6-bis(1,1-dimethylpropyl)-Phenol	25973-55-1	Industrial Chemical	3	Fish		
2,4,6-tris(1,1-dimethylethyl)-Phenol	732-26-3	Industrial Chemical	3	Fish		
3-tert-butyl-4-hydroxy anisole	121-00-6	Industrial Chemical	1	Daphnid		
pentachloro-Benzenethiol	133-49-3	Industrial Chemical	3	Fish		
Pentachlorophenol	87-86-5	Industrial Chemical	1, 2, 3	Fish	Yes	
17α-estradiol	57-91-0	Natural Hormone/Steroid	1	Fish	Yes	Endocrine disruptor (estrogen)
17β-estradiol	50-28-2	Natural Hormone/Steroid	1	Fish	Yes	Endocrine disruptor (estrogen)
Campesterol	474-62-4	Natural Hormone/Steroid	1, 2, 3	Fish		Lowers cholesterol absorption in intestine
Cholestanol	80-97-7	Natural Hormone/Steroid	1	Fish		
Cholesterol	57-88-5	Natural Hormone/Steroid	1, 2, 3	Fish		
Coprostanol	360-68-9	Natural Hormone/Steroid	1, 2, 3	Fish		
Desmosterol	313-04-2	Natural Hormone/Steroid	1, 2, 3	Fish		
Epicoprostanol	516-92-7	Natural Hormone/Steroid	1, 2, 3	Fish		
Equilenin	517-09-9	Natural Hormone/Steroid	1	Fish	Yes	Endocrine disruptor (estrogen)
Ergosterol	57-87-4	Natural Hormone/Steroid	3	Fish		
Estriol	50-27-1	Natural Hormone/Steroid	1	Daphnid	Yes	Endocrine disruptor (estrogen)
Estrone	53-16-7	Natural Hormone/Steroid	1	Daphnid	Yes	Endocrine disruptor (estrogen)
Stigmastanol	19466-47-8	Natural Hormone/Steroid	1, 2, 3	Fish		
Stigmasterol	83-48-7	Natural Hormone/Steroid	1, 2, 3	Fish		

Table A-1. Continued.

Compound	CAS Number	Trace Organic Type	Approach(es) Considered High Priority	Most Sensitive Species (ECOSAR)	Known EDC	Mode of Action or Class of Compound
Testosterone	58-22-0	Natural Hormone/Steroid	1	Daphnid	Yes	Endocrine disruptor (androgen)
β-sitosterol	83-46-5	Natural Hormone/Steroid	1, 2, 3	Fish		
1,3,6,8-tetrabromo-Pyrene	128-63-2	PAH	3	Fish		Carcinogen, gene mutagen
2-Methylfluoranthene	33543-31-6	PAH	3	Fish		Carcinogen, gene mutagen
3-methylcholanthrene	56-49-5	PAH	2, 3	Fish		Carcinogen, gene mutagen
Benzo[a]anthracene	56-55-3	PAH	3	Fish		Carcinogen, gene mutagen
Benzo[a]pyrene	50-32-8	PAH	2, 3	Fish	Yes	Endocrine disruptor (androgen), carcinogen
Benzo[b]fluoranthene	205-99-2	PAH	3	Fish		Immunosuppressant;
Benzo[b]naphtho[2,1-d]thiophene	239-35-0	PAH	3	Fish		Carcinogen, gene mutagen
Benzo[e]pyrene	192-97-2	PAH	3	Fish		Carcinogenic
Benzo[g,h,i]perylene	191-24-2	PAH	3	Fish		Carcinogen, gene mutagen
Benzo[k]fluoranthene	207-08-9	PAH	3	Fish		Carcinogen, gene mutagen
Chrysene	218-01-9	PAH	3	Fish		Carcinogen, gene mutagen
Dibenz[a,h]anthracene	53-70-3	PAH	3	Fish		Carcinogen, gene mutagen
Fluoranthene	206-44-0	PAH	2, 3	Fish		Carcinogen, gene mutagen
Indeno[1,2,3-c,d]pyrene	193-39-5	PAH	3	Fish		Carcinogen, gene mutagen
Perylene	198-55-0	PAH	3	Fish		Carcinogen, gene mutagen
1,1,1,5,5,5-Hexamethyl-3,3-bis[(trimethylsilyl) oxy]-trisiloxane	3555-47-3	Personal Care Product	3	Fish		
1,1,1,5,5,5-hexamethyl-3-phenyl-3-[(trimethylsilyl)oxy]-Trisiloxane	2116-84-9	Personal Care Product	3	Fish		
2,4,6,8-tetraethenyl-2,4,6,8-tetramethyl-Cyclotetrasiloxane	2554-06-5	Personal Care Product	3	Fish		

Table A-1. Continued.

Compound	CAS Number	Trace Organic Type	Approach(es) Considered High Priority	Most Sensitive Species (ECOSAR)	Known EDC	Mode of Action or Class of Compound
2,4,6-trimethyl-2,4,6-tris(3,3,3-trifluoropropyl)-Cyclotrisiloxane	2374-14-3	Personal Care Product	3	Fish		
Cyclohexasiloxane, dodecamethyl-Cyclohexasiloxane	540-97-6	Personal Care Product	3	Fish		
Dodecylmethylpentasiloxane	141-63-9	Personal Care Product	3	Fish		
Heptamethylphenylcyclotetrasiloxane	10448-09-6	Personal Care Product	3	Fish		
Hexadecamethyl heptasiloxane	541-01-5	Personal Care Product	3	Fish		
octadecamethyl-Octasiloxane	556-69-4	Personal Care Product	3	Fish		
Triclosan	3380-34-5	Personal Care Product	1, 2	Fish		
Aldrin	309-00-2	Pesticide	3	Fish	Yes	Endocrine Disruptor (estrogen); CNS stimulation
Benfluralin	1861-40-1	Pesticide	2, 3	Fish		Mitotic disruptor (inhibits plant growth).
Chlorpyrifos	2921-88-2	Pesticide	3	Fish		Neurotoxin
cis-Chlordane	5103-71-9	Pesticide	2, 3	Fish	Yes	Central nervous system stimulation
cis-Nonachlor	5103-73-1	Pesticide	2, 3	Fish		
cis-permethrin	61949-76-6	Pesticide	3	Fish	Yes	Neurotoxin
Dieldrin	60-57-1	Pesticide	2, 3	Fish	Yes	Endocrine disruptor (estrogen); Central nervous system stimulation
Endrin	72-20-8	Pesticide	2, 3	Fish		Central nervous system stimulation
Ethalfluralin	55283-68-6	Pesticide	2, 3	Fish		Inhibits plant cell division
Heptachlor	76-44-8	Pesticide	2, 3	Fish	Yes	Central nervous system stimulation

Table A-1. Continued.

Compound	CAS Number	Trace Organic Type	Approach(es) Considered High Priority	Most Sensitive Species (ECOSAR)	Known EDC	Mode of Action or Class of Compound
Heptachlor epoxide	1024-57-3	Pesticide	3	Fish	Yes	Endocrine disruptor (thyroid)
Hexachlorobenzene	118-74-1	Pesticide	2, 3	Fish	Yes	Endocrine disruptor (thyroid)
Methoxychlor	72-43-5	Pesticide	2, 3	Fish	Yes	Endocrine disruptor (Estrogen); Central nervous system stimulation
Mirex	2385-85-5	Pesticide	2, 3	Fish	Yes	Endocrine disruption (androgen, LH, thyroid?); CNS stimulation
o,p'-DDD	53-19-0	Pesticide	2, 3	Fish	Yes	Endocrine disruption
o,p'-DDE	3424-82-6	Pesticide	2, 3	Fish	Yes	Endocrine disruption
o,p'-DDT	789-02-6	Pesticide	2, 3	Fish	Yes	Endocrine disruptor (Estrogen); Central nervous system stimulation
Octachlorostyrene	29082-74-4	Pesticide	3	Fish	Yes	Endocrine disruption (thyroid)
Oryzalin	19044-88-3	Pesticide	1, 2	Daphnid	Yes	Endocrine disruptor (thyroid); controls weeds by disrupting growth processes during seed germination.
Oxychlordane	27304-13-8	Pesticide	2, 3	Fish	Yes	Endocrine disruption (reproduction); Central nervous system stimulation
p,p'-DDD	72-54-8	Pesticide	2, 3	Fish		Endocrine disruption
p,p'-DDE	72-55-9	Pesticide	2, 3	Fish	Yes	Endocrine disruptor (Androgen)
p,p'-DDT	50-29-3	Pesticide	2, 3	Fish	Yes	Endocrine disruptor (Estrogen); Central nervous system stimulation
Pendimethalin	40487-42-1	Pesticide	3	Fish	Yes	Endocrine disruptor (Thyroid)

Table A-1. Continued.

Compound	CAS Number	Trace Organic Type	Approach(es) Considered High Priority	Most Sensitive Species (ECOSAR)	Known EDC	Mode of Action or Class of Compound
Pentachloroanisole (PCA)	1825-21-4	Pesticide	2, 3	Fish		Fungicide (metabolite of pentachlorophenol)
Permethrin	52645-53-1	Pesticide	3	Fish	Yes	Endocrine disruption (estrogen); Neurotoxin
Perthane	72-56-0	Pesticide	3	Fish		Organochlorine insecticide
Propargite	2312-35-8	Pesticide	3	Fish		
trans-Chlordane	5103-74-2	Pesticide	2, 3	Fish	Yes	Central nervous system stimulation
trans-Nonachlor	39765-80-5	Pesticide	2, 3	Fish	Yes	Endocrine disruptor (Estrogen); Central nervous system stimulation
trans-permethrin	61949-77-7	Pesticide	3	Fish	Yes	Neurotoxin
Trifluralin	1582-09-8	Pesticide	2, 3	Fish	Yes	Endocrine disruptor (reproduction; metabolism); meristematic inhibitor in plants (stops plant cells from dividing)
γ-Chlordane	5566-34-7	Pesticide	2, 3	Fish		Central nervous system stimulation
17α-ethynylestradiol	57-63-6	Pharmaceutical	1, 2, 3	Daphnid	Yes	Endocrine disruptor (estrogen)
Clotrimazole	23593-75-1	Pharmaceutical	2, 3	Daphnid		Antifungal
Fenoprofen	31879-05-7	Pharmaceutical	1	Fish		Cycoloxygenase (COX) Inhibition
Lofepramine	23047-25-8	Pharmaceutical	3	Fish		Inhibits norepinephrine neuronal transport
Mestranol	72-33-3	Pharmaceutical	1, 2, 3	Fish	Yes	Endocrine Disruptor (estrogen)
Miconazole	22916-47-8	Pharmaceutical	2, 3	Fish		Antifungal agent
Norethisterone	68-22-4	Pharmaceutical	1	Daphnid	Yes	Endocrine disruptor (estrogen)

Table A-1. Continued.

Compound	CAS Number	Trace Organic Type	Approach(es) Considered High Priority	Most Sensitive Species (ECOSAR)	Known EDC	Mode of Action or Class of Compound
Novobiocin	303-81-1	Pharmaceutical	1	Fish		Inhibits DNA gyrase by targeteing the GyrB subunit
Propranolol	525-66-6	Pharmaceutical	1	Daphnid		Beta adrenergic receptor antagonist
Tamoxifen	10540-29-1	Pharmaceutical	1, 2, 3	Fish	Yes	Endocrine disruptor (estrogen)
β-estradiol-3-benzoate	50-50-0	Pharmaceutical	3	Fish	Yes	Endocrine disruptor (estrogen)
Bis(2-ethylhexyl)adipate	103-23-1	Plasticizer	1, 2, 3	Fish	Yes	
Bis(2-ethylhexyl)phthalate	117-81-7	Plasticizer	1, 2, 3	Fish	Yes	Endocrine disruptor (androgen)
Bisphenol A	80-05-7	Plasticizer	1	Daphnid	Yes	Endocrine disruptor (estrogen)
Di-N-octyl phthalate	117-84-0	Plasticizer	1, 2, 3	Fish	Yes	Endocrine disruptor (estrogen)
4-n-nonylphenol	104-40-5	Surfactant	1, 2, 3	Fish	Yes	Endocrine disruptor (estrogen)
4-n-octylphenol	1806-26-4	Surfactant	3	Daphnid	Yes	Endocrine disruptor (estrogen)
4-Nonylphenol	25154-52-3	Surfactant	1, 2, 3	Fish	Yes	Endocrine disruptor (estrogen)
4-nonylphenol diethoxycarboxylate	106807-78-7	Surfactant	1, 2	Fish	Yes	Endocrine disruptor (estrogen)
4-nonylphenol monoethoxycarboxylate	3115-49-9	Surfactant	1, 2	Fish	Yes	Endocrine disruptor (estrogen)
4-nonylphenol monoethoxylate	27986-36-3	Surfactant	1, 2	Fish	Yes	Endocrine disruptor (estrogenic)
4-tert-octylphenol	140-66-9	Surfactant	3	Fish	Yes	ER agonist; reduces estradiol binding to ER
Octylphenol	27193-28-8	Surfactant	3	Fish	Yes	Endocrine disruptor (estrogen)

Table A-1. Continued.

Compound	CAS Number	Trace Organic Type	Approach(es) Considered High Priority	Most Sensitive Species (ECOSAR)	Known EDC	Mode of Action or Class of Compound
para-nonylphenol	84852-15-3	Surfactant	1, 2, 3	Fish		
3-chloro-2,6-dinitro-N,N-dipropyl-4-(trifluoromethyl)-Benzenamine	29091-20-1	Unknown	3	Fish		
Bicyclo 2.2.1 hept-5-ene-2,3-dicarboxylic acid, 1,4,5,6,7,7-hexachloro-, dibutyl ester	1770-80-5	Unknown	3	Fish		
tetrabromo-, bis(2-ethylhexyl) ester Phthalic acid	26040-51-7	Unknown	3	Fish		
tetracosafluorotetradecahydro-Phenanthrene	306-91-2	Unknown	3	Fish		
triphenyl ester Phosphorous acid	101-02-0	Unknown	3	Fish		

Table A-2. Analytical Methods and Their Associated Quantitation Limits for TOrCs Identified as High Priority Using Any of the Three Prioritization Approaches in Task 1 of this Research. Cells with " - " indicate that information was not available or readily obtained. Numbers in [ ] refer to references at the end of this table. Superscript letters refer to footnotes at the end of this table.

TOrC	Method	MDL(ng/L)[a]	MRL(ng/L)[a]
o,p'-DDD	[1]	3.5	11
p,p'-DDD	[2][b] [3]	 25	100 83
o,p'-DDE	[2]	0.5	1.7
p,p'-DDE	[2][b] [3]	 1	100 3.3
o,p'-DDT	[2]	21	69
p,p'-DDT	[2][b] [3]	 18	100 58
OTNE	[4]		4
PBDE-209	[5]	1	10[c]
Acetyl cedrene	[4]		7
Acetyl-hexamethyl-tetrahydro-napthalene (AHTM)	NA		
Aldrin	[2][b]		40
Benfluralin	[6]		50
Benzo[a]anthracene	[2][b]		50
Benzo[g,h,i]perylene	[2][b]		50
2-(2H-benzotriazol-2-yl)-4,6-bis(1,1-dimethylpropyl)-Phenol	NA		
Bis(2-ethylhexyl)adipate	[2][b]		600
Bis(2-ethylhexyl)phthalate	[2][b]		600
Bisphenol A	[7] [2] [8]	 5.4 	1. 5 18 30
3-tert-butyl-4-hydroxy anisole	NA		
Bicyclo 2.2.1 hept-5-ene-2,3-dicarboxylic acid, 1,4,5,6,7,7-hexachloro-, dibutyl ester	NA		

Table A-2. Continued.

Benzo[b]fluoranthene	[2][b]		20
Benzo[k]fluoranthene	[2][b]		20
Benzo[b]naphtho[2,1-d]thiophene	NA		
Benzo[a]pyrene	[2][b]		20
Benzo[e]pyrene	[2][b]		20
1,1'-Biphenyl, bis(1-methylethyl)- 1,1,1,5,5,5-hexamethyl-3-phenyl-3- [(trimethylsilyl)oxy]-Trisiloxane	18-20[e]		
Campesterol	[9]	4.6	10
Celestolide (ADBI)	NA		
cis-Chlordane	[2][b]		50
trans-Chlordane	[2]	220	
γ-Chlordane	[2][b]		50
3-chloro-2,6-dinitro-N,N-dipropyl -4-(trifluoromethyl)-Benzenamine	NA		
Chlorpyrifos	[2][b] [7]		100 50
Cholestanol	[9]	2.7	10
Cholesterol	[9]	0.6 (estimated)	10
Chrysene	[2][b]		20
Clotrimazole	NA		
Coprostanol	[9]	1.3	2
Cyclohexasiloxane	18-20[e]		
dodecamethyl-cyclohexasiloxane	18-20		
Desmosterol	[9]	5.2	20
Dieldrin	[2][b]		200
Dibenz[a,h]anthracene	[2][b]		50
2,4,6-tris(1,1-dimethylethyl)-Phenol	NA		
1,3-DiNO$_2$ 2-Cl 5-CF3 benzene	NA		

Table A-2. Continued.

Di-N-octyl phthalate	NA			
Dodecylmethylpentasiloxane	18-20			
Endrin	[2][b]		100	
Epicoprostanol	[9]	0.7	2	
Equilenin	[10]	0.1	2	
Ergosterol	[9]	20 (estimated)	50	
17α-estradiol	[9]	0.1	2	
	[8]		10	
17β-estradiol	[9]	0.1	2	
	[10]		<1	
	[7]		0.5	
	[8]		10	
β-estradiol-3-benzoate	[9]	0.6	2	
Estriol	[10]	0.1	2	
	[10]		<1	
	[8]		10	
Estrone	[9]	0.2	2	
	[10]			1-10
	[7]		0.2	
	[8]		10	
Ethalfluralin	[6]		50	
1,1'- 1,2-ethanediyl bis(oxy) bis 2,4,6-tribromo-Benzene	NA			
1,1'-ethylidenebis 3,4-dimethyl-Benzene	NA			
Fenoprofen	NA			
17α-Ethynylestradiol	[9]	0.1	2	
	[10]		1-3	
	[7]		1	
	[8]		10	
Fluoranthene	[2][b]		100	
Galaxolide	[11]		5000	
Heptachlor	[2][b]		40	
Heptachlor epoxide (isomer A)	[2][b]		100	
Heptachlor epoxide (isomer B)	[2][b]		20	

Table A-2. Continued.

Heptamethylphenylcyclotetrasiloxane	18-20[e]		
Hexabromocyclododecane	NA		
Hexachlorobenzene	[2][b]		50
Hexadecamethyl heptasiloxane	18-20[e]		
1,1,1,5,5,5-Hexamethyl- 3,3-bis[(trimethylsilyl) oxy]-trisiloxane	18-20[e]		
Indeno[1,2,3-c,d]pyrene	[2][b]		50
Lofepramine	NA		
Mestranol	[9]	0.1	2
Methoxychlor	[2][b]		100
3-methylcholanthrene	NA		
2-Methylfluoranthene	NA		
Miconazole	[12]	1.3	5
Mirex	[13] [14]	22 0.035	
Musk ketone	[7] [4]		21 0.7
Musk xylene	[4]		1
Napthylene	[2][b]		100
cis-Nonachlor	[2][b]		
trans-Nonachlor	[2][b]		50
Nonylphenol	[7] [3]	 7.5	80[c] 25[c]
4-Nonylphenol	[7]		1000
4-n-nonylphenol	[3]	67	220
para-nonylphenol	[3]		
Nonylphenol diethoxycarboxylate	[15]	15	
Nonylphenol monoethoxycarboxylate	[15]	2.5	
Nonylphenol monoethoxylate	[3] [8]	4.9	16 10

Table A-2. Continued.

Nonylphenol diethoxylate	[3] [8]	0.7	2.3 10
Norethisterone	NA		
Novobiocin	NA		
Octachlorostyrene	NA		
octadecamethyl-Octasiloxane	18-20[e]		
Octylphenol	[7] [3]	 19	21[c] 64[c]
4-n-Octylphenol[13]	[8]		1000
Octylphenol monoethoxylate[13]	[3]	0.7	2.3
Octylphenol diethoxylate[13]	[3]	39	130
4-n-octylphenol	[3]	220	730
4-tert-octylphenol	[3]		
Oryzalin	[6] [7]		50 50
Oxychlordane	NA		
Pendimethalin	NA		
Pentachloroanisole (PCA)	NA		
pentachloro-Benzenethiol	NA		
Pentachlorophenol	[2][b] [8]		1000 1000
cis-permethrin	[2]		34
trans-permethrin	[2]		67
Permethrin (mixed isomers)	[2][b]		100
Perthane	NA		
Perylene	[16]	1.4	
Phantolide (AHMI)	NA		
Phenanthrene	[2][b]		20
Propargite	NA		

Table A-2. Continued.

Propranolol	[1]	0.6	2.1
β-sitosterol	[9]	1.2est	6
Stigmastanol	[9], [10]	2.0est	6
Stigmasterol	[8]	0.6est	2
Tamoxifen	[17]	6-15	
Testosterone	[9]	0.7	2
	[10]		0.1-1
	[1]	1.7	5.8
	[7]		0.5
	[8][d]		
tetrabromo-, bis(2-ethylhexyl) ester phthalic acid	NA		
1,3,6,8-tetrabromo-Pyrene	NA		
tetracosafluorotetradecahydro-Phenanthrene	NA		
2,4,6,8-tetraethenyl-2,4,6,8-tetramethyl-cyclotetrasiloxane	18-20[e]		
Tonalide	NA		
Traseolide (ATII)	NA		
Triclosan	[10]	92	200
	[6]		1
	[7]		1
Trifluralin	[2][b]		100
	[5]		50
2,4,6-trimethyl-2,4,6-tris (3,3,3-trifluoropropyl)-Cyclotrisiloxane	18-20[e]		
Triphenyl ester Phosphorous acid	NA		

Footnotes

a. MDL is the Minimum Detection Limit; MRL is the Minimum Reporting Level which is defined differently depending on the method.
b. MWH Laboratories Minimum Reporting Levels for analytes in EPA Method 525.2. Minimum detection limits are not presented for these analytes
c. Reported by Columbia Analytical Services (http://www.caslab.com/Polybrominated-Diphenyl-Ethers-Testing/)
c. Reported as total nonylphenol and octylphenol in reference [6].
d. Reference specifies trans-Testosterone
e. In principle, the methods used in references 18-20 should be amenable to analysis of these compounds

Table A-2. Continued.

References for Table 2-4

1. A. Batt, M. Kostich, and J. Lazorchak, "Analysis of Ecologically Relevant Pharmaceuticals in Wastewater and Surface Water Using Selective Solid-Phase Extraction and UPLC-MS/MS," Anal. Chem., **80**, 5021, 2008.

2. U.S. EPA Method 525.2, Determination of Organic Compounds in Drinking Water by Liquid-Solid Extraction and Capillary Column Gas Chromatography/Mass Spectrometry, Revision 2.0, National Exposure Research Laboratory, 1995.

3. J-P. Baugros, B. Giroud, G. Dessalces, M-F Grenier-Loustalot, and C. Cren-Olive, Multiresidue, Analytical Methods, for the Ultra-Trace Quantification of 33 Priority Substances Present in the List of REACH in Real Water Sample, Analytica Chimica Acta, **607**, 191, 2008.

4. S. Simonich, W. Begley, G. Debaere, and W. Eckhoff, "Trace Analysis of Fragrance Materials in Wastewater and Treated Wastewater," Environ. Sci. Technol., 34, 959, 2000.

5. U.S. EPA 8270C, Semivolatile Organic Compounds by Gas Chromatography/Mass Spectrometry (GC/MS).

6. California Department of Food and Agriculture, "Determination of Ethalfluralin, Trifluralin, Benfluralin, Prodiamine, Pendimethalin, Oxyfluorfen, and Oryzalin in Surface Water," September 2007.

7. These detection and reporting limits are from the following references: B Vanderford and S. Snyder, "Analysis of Pharmaceuticals in Water by Isotope Dilution Liquid Chromatography/Tandem Mass Spectrometry" Environ. Sci. Tech., **40**, 7312, 2005; R. Lavado et al., "Site-Specific Profiles of Estrogenic Activity in Agricultural Areas of California's Inland Waters," Environ. Sci Technol., 43, 9110, 2009; M. Benotti et, al., "Pharmaceuticals and Endocrine Disrupting Compounds in U.S. Drinking Water," Environ. Sci. Technol., **43**, 597, 2009.

8. Y. Gou, S. Krasner, S, Fitzsimmons, G. Woodside, and N. Yamachika, "National Water Research Institute Final Project Report: Source, Fate, and Transport of Endocrine Disruptors, Pharmaceuticals, and Personal Care Products in Drinking Water Sources in California," May 2010.

9. U.S. EPA Methods 1698, Steroids and Hormones in Water, Soil, Sediment, and Biosolids by HRGC/HRMS, Office of Science and Technology, 200710. USEPA Method 1694, Pharmaceuticals and Personal Care Products in Water, Soil, Sediment, and Biosolids by HPLC/MS/MS, Office of Science and Technology, 2007.

10. U.S. EPA, Method 539, Method Under Review, Office of Ground Water and Drinking Water, 2010. Minimum reporting limits presented as a <value or in a range, since method is under review.

11. W. Schüssler and L.Nitschke. 1997. Determination of trace amounts of Galaxolide® (HHCB) by HPLC. Fresenius' Journal of Analytical Chemistry 361: 220-221.

12. U.S. EPA Method 1694, Pharmaceuticals and Personal Care Products in Water, Soil, Sediment, and Biosolids by HPLC/MS/MS, Office of Science and Technology, 2007.

13. U.S. EPA. Method 527: Pesticides and flame retardants in water by SPE and capillary column GC/MS.

14. U.S. EPA Method 1699: Pesticides in water, soil, sediment, biosolids, and tissue by HRGC/HRMS.

15. Y. Koh, T. Chiu, A. Boobis, E. Cartmell, S. Pollard, M. Scrimshaw, and J. Lester, "A Sensitive and Robust Method for the Determination of Alkylyphenol Polyethoxylates and their Carboxylic Acids and their Transformation in a Trickling Filter Wastewater Treatment Plant, Chemisphere, **73**, 551, 2008.

16. P, Crozier, J. Plomley, L. Matchuk, "Trace Level Analysis of Polycyclic Aromatic Hydrocarbons in Surface Waters by Solid Phase Extraction (SPE) and Gas Chromatography-Ion Trap Mass Spectrometry, Analyst, **126**, 1974, 2001.

Table A-2. Continued.

17. A. Tauxe-Wuersch, L. DeAlencastro, D. Grandjean, and J. Tarradellas, "Trace Determination of Tamoxifen and 5-Fluorouracil in Hospital and Urban Wastewaters," International Journal of Environmental Analytical Chemistry, 86, 473, 2006.

18. E. Brorström-Lundén, L.Kaj, J. Andersson, A. Palm Cousins, and M. Remberger, IVL, Swedish Environmental Research Institute and M. Schlabach and N. Schmidbauer, NILU, Norwegian Institute for Air Research, " Screening of New Chemicals in Sweden: The Siloxane Case," Presentation, 2004.

19. L.Kaj, J. Andersson, A. Palm Cousins, M. Remberger, E. Brorström-Lundén and I. Cato, SGU, Geological Survey of Sweden, "Results from the Swedish National Screening Programme, Subpart 4:Siloxanes," 2004.

20. Norman Bulletin on Emerging Substances, Norman Network of Reference Laboratories for Monitoring of Environmental Substances, Issue 1, 2009 – http://www.norman-network.net.

APPENDIX B

WERF TRACE ORGANIC COMPOUNDS (TORCS) DATABASE DATA DICTIONARY

WERF Trace Organic Compounds (TOrCs) Database Data Dictionary

Version 1.2

October 21, 2010

Prepared for
Water Environment Research Foundation (WERF)

Prepared by
Tetra Tech, Inc.
Fairfax, VA 22030

Owings Mills, MD 21157

WERF
Water Environment Research Foundation
Collaboration. Innovation. Results.

Revision History

Table 1: Change Record.

Version Number	Description of Change	Change Effective Date	Change Entered By
1.0	Initial Draft	11/16/2009	Tetra Tech, Inc.
1.1	Data Dictionary Review, Edit, & Comment	12/01/2009	Tetra Tech, Inc.
1.2	Data Dictionary Review, Edit, & Comment	10/21/2010	Tetra Tech, Inc.

Table of Contents

1.0 Introduction

1.1 Purpose and Scope

This section provides a brief description of the Systems Design Document's purpose and scope.

1.1.1 System Overview

The core requirements for the tool are:

1. **Data Entry**: Although the tool will initially be loaded with data from selected studies, stakeholders require the ability to update the database with results collected during subsequent analyses. Users will require the ability to manually add (and edit) information associated with studies (reference/citation), locations, and results.

2. **Data Upload**: Stakeholders working with larger projects require the ability to batch load data from their internal data systems, rather than manually enter the values.

3. **Reporting**: Stakeholders require the ability to generate reports summarizing the results from selected locations and chemical compounds. Additionally, users may wish to compare selected results with national averages or other pre-defined geographic ranges.

4. **Mapping**: Users require the ability to view maps showing locations where monitoring was conducted. Maps might also be used to select areas of interest for report generation.

2.0 Data Dictionary

2.1 Data Element Dictionary

This section presents the WERF data element dictionary that has been captured in two reports:

- The Entity Report lists all the tables in the WERF database and describes the function of each table.
- The Data Dictionary Report contains a list of all the entities and attributes in the WERF database the definition of each attribute the entity/table in which it is stored and several columns of information about the attribute. Both reports are printed alphabetically by table name.

Table 2-1. Naming Conventions Used to Describe WERF Data.

Common Names	Convention
Primary Key Names	The system generated primary key identifying unique records in each element will be named as "ELEMENT_ID". For example, the primary key for the STUDY table will be named STUDY_ID. Foreign key references will also be identified by ending with "_ID".
Look up value	Tables containing look up values will be prefaced with "LV_" (for example, LV_COUNTY).
"Yes" or "No" select options	Table attributes storing "Yes/No" information will be named with a "_YN" following the field name
Created/Modified By and Date	Each element will contain attributes identifying the user and date that created or edited each record.

2.2 Entity Report

This report contains the following columns of information (Table 2-2):

- Entity Name – is the specific name of each table in the WERF database. The name usually indicates the kind of WERF data that is stored in the table.
- Entity Definition – gives a brief summary description of the kind of data stored in the table and the function of the table. Table functions include providing data reference or other types of information.
  Example: WERF_ADDRESSES table contains the mailing addresses for all the entities.

Table 2-2. Description of Entity Names and Definitions in the WERF Database.

TABLE_NAME	TABLE_DESCRIPTION
ATTACHMENT	Table to store attachments (photos, documents, pdf, etc).
BIOLOGICAL_RESULT	Biological results are meant to store the actual measurements taken during sampling events at a location.
COMPOUND_RESULT	Chemical compound results are meant to store the actual measurements.
EMAIL	Email for users
HABITAT_RESULT	Habitat results are meant to store the actual habitat measurements taken during sampling events at a location.
LOCATION	Locations represent the physical place on the earth where measurements were collected.
LV_ANALYSIS_METHOD	Standard analysis method
LV_ANALYSIS_METHOD_CONTEXT	Context for analysis method
LV_ATTACHMENT_TABLE	Look up containing list of table names that may have an attachment.
LV_BIOLOGICAL_INTENT	Look up options for biological intent such as fish sampling
LV_COMPOUND	This Entity is meant to store all of the compound names (and metadata) for which measurements are recorded.
LV_COUNTRY	Look up values for countries
LV_COUNTY	Look up values for counties
LV_DATA_QUALIFIER	Look up values for qualifiers bracketing results.
LV_DETECTION_LIMIT_TYPE	Look up values for the type of detection limit.
LV_EVENT_TYPE	Look up values for event type.
LV_HABIT	Look up for habit names or descriptors
LV_HABITAT	Look up for habitat names or descriptors
LV_LIFE_STAGE	Look up for life stage of the taxon
LV_LOACTION_TYPE	Look up values for types of locations (lake, stream, and river).
LV_MEASUREMENT_TYPE	Look up values for the type of measurement
LV_MEDIA_SUBDIVISION_TYPE	Look up values for media subtype, (i.e. Wastewater, freshwater)
LV_MEDIA_TYPE	Look up values for media types (i.e., water)

Table 2-2 Continued.

TABLE_NAME	TABLE_DESCRIPTION
LV_SAMPLE_FRACTION	Look up values for sample fraction
LV_SAMPLE_TYPE	Look up descriptions for sample type
LV_STATE	Look up values for states.
LV_STUDY_STATUS	Look up values for Study (accepted / Rejected).
LV_TAXONOMY	Look up for taxonomic names
LV_TAXONOMY_TYPE	Taxonomy type such as fish, benthic macroinvertebrates
LV_TISSUE_ANATOMY	Look up for part of the anatomy where the fish tissue taken
LV_TRACE_ORGANIC_TYPE	Look up values for the type of trace organic for a compound.
LV_TROPHIC_LEVEL	Look up for trophic level of the taxon
LV_UNIT_CONVERSION	Look up values for the conversion factors and
LV_UNIT_TYPE	Look up values for valid types of measurement units in the system.
LV_USE_AND_OR_SECONDARY_TYPE	Look up values for secondary types for a compound
LV_WEIGHT_BASIS	Look up values for the type of weight / basis associated with a sample.
SAMPLING_EVENT	Events represent groups of results that have been collected at a location at a particular time, or measurements that have come from a singular "batch" or media sample.
STUDY	This application includes measurements from a variety of STUDIES. Studies represent a single parent record for all subsequent data (Locations, Events & Results).

2.3 Data Dictionary Report

This report contains the following columns of information (Table 2-3):

- **Entity Name** – is the specific name of each table in the WERF database. The name usually indicates the kind of WERF data that is stored in the table. For entity name standards refer to database overview section in this document.

  Example: The LV_WERF_PERSON table contains data regarding all the people concerned with the WERF.

- **Attribute Name** – contains the name of the field in which data may be stored on the WERF database in a specified table. For attribute name standards refer to database overview section in this document.

  Example: PERS_ID contains the unique identifier of a person.

- **Description** – is an explanation of the data stored under the attribute name

  Example: PERS_ID is a system-generated number that is assigned to uniquely identify a person.

- **Format (i.e. Data Type Size)** – indicates the acceptable format of the data stored under this attribute name.
  Example: VARCHAR2(10) means the field will accept data containing up to 10 numeric and/or alphabetic characters DATE means the field will accept only valid calendar dates and NUMBER(5) means the field will accept only a numerical entry with up to five digits.

- Length – indicates how long the character string or size of the number that can be stored in the field.
- **Required** – Can the field be left blank or is a value required.

- **Primary Key** – does this field serve as a primary key for the table, meaning it cannot be blank and must be unique for each record.

- **Foreign Key** – does this field act as a link to another database table.

- **Reference Table** – does this table reference another table through a foreign key, usually a look-up table.

Table 2-3. WERF Data Dictionary Report.

TABLE_NAME	COLUMN_NAME	DESCRIPTION	DATA TYPE	LENGTH	REQUIRED	PRIMARY KEY	FOREIGN KEY	REFERENCE TABLE
ATTACHMENT	ATTACHMENT_ID	Database generated unique ID (record identifier) for the attachment	INT	0	NOT NULL	PRIMARY KEY		
ATTACHMENT	ATTACHMENT_TABLE_ID	ID from the look up table for attachments	INT	0	NOT NULL		FOREIGN KEY	LV ATTACHMENT TABLE
ATTACHMENT	ENTITY_ID	Database entity associated with attachment	INT	0	NOT NULL			
ATTACHMENT	NAME	Name given to the attachment by the user	VARCHAR	100	NOT NULL			
ATTACHMENT	DESCRIPTION	Description of the attachment	VARCHAR	-1	NULL			
ATTACHMENT	FILE_DATA	Stores an image	IMAGE	2E+09	NOT NULL			
ATTACHMENT	CREATED_BY	UserID of person that added the attachment	INT	0	NOT NULL			
ATTACHMENT	CREATED_DATE	Date attachment was added	VARCHAR	50	NOT NULL			
ATTACHMENT	MODIFIED_BY	UserID of person that modified the attachment	INT	0	NULL			
ATTACHMENT	MODIFIED_DATE	Date attachment was modified	VARCHAR	50	NULL			
BIOLOGICAL_RESULT	BIOLOGICAL_RESULT_ID	Database generated unique ID (record identifier)	INT	0	NOT NULL	PRIMARY KEY		

Table 2-3. Continued.

TABLE_NAME	COLUMN_NAME	DESCRIPTION	DATA TYPE	LENGTH	REQUIRED	PRIMARY KEY	FOREIGN KEY	REFERENCE TABLE
BIOLOGICAL_RESULT	SAMPLING_EVENT_ID	ID from the SAMPLING_EVENT table that identifies the sample	INT	0	NOT NULL		FOREIGN KEY	SAMPLING EVENT
BIOLOGICAL_RESULT	TAXONOMY_ID	ID from the LV_TAXONOMY table that identifies the Taxon	INT	0	NOT NULL		FOREIGN KEY	LV TAXONOMY
BIOLOGICAL_RESULT	BIOLOGICAL_INTENT_ID	ID from the LV_BIOLOGICAL_INTENT that identifies the biological intent	INT	0	NULL		FOREIGN KEY	LV BIOLOGICAL INTENT
BIOLOGICAL_RESULT	LIFE_STAGE_ID	ID that identifies the life stage of the biological organism (TAXON)	INT	0	NULL		FOREIGN KEY	LV LIFE STAGE
BIOLOGICAL_RESULT	TROPHIC_LEVEL_ID	ID that identifies the trophic level of the biological organism (TAXON)	INT	0	NULL		FOREIGN KEY	LV TROPHIC LEVEL
BIOLOGICAL_RESULT	HABIT_ID	ID that identifies the Habit of the biological organism (TAXON)	INT	0	NULL		FOREIGN KEY	LV HABIT
BIOLOGICAL_RESULT	COUNT	The count or number of biological organisms in the sample	INT	0	NOT NULL			

Table 2-3. Continued.

TABLE_NAME	COLUMN_NAME	DESCRIPTION	DATA TYPE	LENGTH	REQUIRED	PRIMARY KEY	FOREIGN KEY	REFERENCE TABLE
BIOLOGICAL_RESULT	POLLUTION_TOLERANCE	The pollution tolerance rating number or category of the biological organism	VARCHAR	30	NULL			
BIOLOGICAL_RESULT	POLLUTION_TOLERANCE_SCALE	The rating scale of the pollution tolerance value given in the "Pollution Tolerance"	VARCHAR	30	NULL			
BIOLOGICAL_RESULT	EXCLUDE_YN	Y indicates the biological organism should be excluded from metric calculations	BIT	0	NOT NULL			
BIOLOGICAL_RESULT	COMMENTS	Free text comments on a biological result	VARCHAR	-1	NULL			
BIOLOGICAL_RESULT	CREATED_DATE	System generated date of when the record was created	DATETIME	0	NOT NULL			
BIOLOGICAL_RESULT	CREATED_BY	User ID of who created the record	VARCHAR	50	NOT NULL			
BIOLOGICAL_RESULT	MODIFIED_DATE	System generated date of when the record was modified	DATETIME	0	NULL			
BIOLOGICAL_RESULT	MODIFIED_BY	User ID of who modified the record	VARCHAR	50	NULL			

Table 2-3. Continued.

TABLE_NAME	COLUMN_NAME	DESCRIPTION	DATA TYPE	LENGTH	REQUIRED	PRIMARY KEY	FOREIGN KEY	REFERENCE TABLE
COMPOUND_RESULT	COMPOUND_RESULT_ID	Database generated unique ID (record identifier)	INT	0	NOT NULL	PRIMARY KEY		
COMPOUND_RESULT	SAMPLING_EVENT_ID	ID from the SAMPLING_EVENT table that identifies the sample	INT	0	NOT NULL		FOREIGN KEY	SAMPLING_E VENT
COMPOUND_RESULT	COMPOUND_ID	ID from the LV_COMPOUND table that identifies the compound	INT	0	NOT NULL		FOREIGN KEY	LV COMPOUND
COMPOUND_RESULT	RESULT_VALUE	The measure result or value for the compound	NUMERIC	0	NULL			
COMPOUND_RESULT	RESULT_VALUE_UNIT_TYPE_ID	ID from the LV_UNIT_TYPE table that identifies the measurement unit	INT	0	NULL		FOREIGN KEY	LV UNIT TYPE
COMPOUND_RESULT	ORIGINAL_RESULT_VALUE	The measured result or value of the compound before being transformed to a standard unit	NUMERIC	0	NULL			
COMPOUND_RESULT	ORIGINAL_RESULT_VALUE_UNIT_TYPE_ID	The original measurement unit before being transformed to a standard unit	INT	0	NULL		FOREIGN KEY	LV UNIT TYPE

Table 2-3. Continued.

TABLE_NAME	COLUMN_NAME	DESCRIPTION	DATA TYPE	LENGTH	REQUIRED	PRIMARY KEY	FOREIGN KEY	REFERENCE TABLE
COMPOUND_RESULT	MEASUREMENT_TYPE_ID	ID from the LV_MEASUREMENT_TYPE table that identifies the type of measurement	INT	0	NOT NULL		FOREIGN KEY	LV MEASUREMENT TYPE
COMPOUND_RESULT	TRACE_ORGANIC_TYPE_ID	ID from the LV_TRACE_ORGANIC_TYPE table that identifies the type of trace organic	INT	0	NULL		FOREIGN KEY	LV TRACE ORGANIC TYPE
COMPOUND_RESULT	USE_AND_OR_SECONDARY_TYPE_ID	ID from the LV_USE_AND_OR_SECONDARY_TYPE table that identifies the use of or secondary type of compound	INT	0	NULL		FOREIGN KEY	LV USE AND OR SECONDARY TYPE
COMPOUND_RESULT	MEDIA_TYPE_ID	ID from the LV_MEDIA_TYPE table that identifies the type of media	INT	0	NOT NULL		FOREIGN KEY	LV MEDIA TYPE
COMPOUND_RESULT	MEDIA_SUBDIVISION_TYPE_ID	ID from the LV_MEDIA_SUBDIVISION_TYPE table that identifies the subdivision for the media	INT	0	NULL		FOREIGN KEY	LV MEDIA SUBDIVISION TYPE
COMPOUND_RESULT	DATA_QUALIFIER_ID	ID from the LV_DATA_QUALIFIER table that identifies the data qualifier	INT	0	NULL		FOREIGN KEY	LV DATA QUALIFIER

Table 2-3. Continued.

TABLE_NAME	COLUMN_NAME	DESCRIPTION	DATA TYPE	LENGTH	REQUIRED	PRIMARY KEY	FOREIGN KEY	REFERENCE TABLE
COMPOUND_RESULT	ANALYSIS_METHOD_ID	ID from the LV_ANALYSIS_METHOD table that identifies the analysis method	INT	0	NULL		FOREIGN KEY	LV ANALYSIS METHOD
COMPOUND_RESULT	SAMPLE_FRACTION_ID	ID from the LV_SAMPLE_FRACTION table that identifies the sample fraction	INT	0	NULL		FOREIGN KEY	LV SAMPLE FRACTION
COMPOUND_RESULT	WEIGHT_BASIS_ID	ID from the LV_WEIGHT_BASIS table that identifies the weight basis	INT	0	NULL		FOREIGN KEY	LV WEIGHT BASIS
COMPOUND_RESULT	TISSUE_ANATOMY_ID	ID from the LV_TISSUE_ANATOMY table that identifies the anatomy from which the tissue was taken	INT	0	NULL		FOREIGN KEY	LV TISSUE ANATOMY
COMPOUND_RESULT	TAXONOMY_ID	ID from the LV_TAXONOMY table that identifies the Taxon	INT	0	NULL		FOREIGN KEY	LV TAXONOMY
COMPOUND_RESULT	TAXONOMY_COUNT	The count or number of biological organisms in the sample	INT	0	NULL			

Table 2-3. Continued.

TABLE_NAME	COLUMN_NAME	DESCRIPTION	DATA TYPE	LENGTH	REQUIRED	PRIMARY KEY	FOREIGN KEY	REFERENCE TABLE
COMPOUND_RESULT	DETECTION_LIMIT_TYPE_ID	ID from the LV_DETECTION_LIMIT_TYPE table that identifies the type of detection limit	INT	0	NULL		FOREIGN KEY	LV DETECTION LIMIT TYPE
COMPOUND_RESULT	DETECTION_LIMIT_VALUE	The minimum detection limit value	NUMERIC	0	NULL			
COMPOUND_RESULT	DETECTION_LIMIT_VALUE_UNIT_TYPE_ID	ID from the LV_UNIT_TYPE table that identifies the measurement unit	INT	0	NULL		FOREIGN KEY	LV UNIT TYPE
COMPOUND_RESULT	REPORTING_LIMIT_VALUE	The minimum reporting limit value	NUMERIC	0	NULL			
COMPOUND_RESULT	REPORTING_LIMIT_VALUE_UNIT_TYPE_ID	ID from the LV_UNIT_TYPE table that identifies the measurement unit	INT	0	NULL		FOREIGN KEY	LV UNIT TYPE
COMPOUND_RESULT	COMMENTS	Free text comments on a compound result	VARCHAR	-1	NULL			
COMPOUND_RESULT	CREATED_DATE	System generated date of when the record was created	DATETIME	0	NOT NULL			
COMPOUND_RESULT	CREATED_BY	User ID of who created the record	VARCHAR	50	NOT NULL			

Table 2-3. Continued.

TABLE_NAME	COLUMN_NAME	DESCRIPTION	DATA TYPE	LENGTH	REQUIRED	PRIMARY KEY	FOREIGN KEY	REFERENCE TABLE
COMPOUND_RESULT	MODIFIED_DATE	System generated date of when the record was modified	DATETIME	0	NULL			
COMPOUND_RESULT	MODIFIED_BY	User ID of who modified the record	VARCHAR	50	NULL			
EMAIL	EMAIL_ID	Database generated unique ID (record identifier)	INT	0	NOT NULL	PRIMARY KEY		
EMAIL	OWNER_NAME	User name for email address	VARCHAR	50	NOT NULL			
EMAIL	EMAIL_ADDRESS	The user's email address	VARCHAR	50	NOT NULL			
EMAIL	CREATED_DATE	System generated date of when the record was created	DATETIME	0	NOT NULL			
EMAIL	CREATED_BY	User ID of who created the record	VARCHAR	50	NOT NULL			
EMAIL	MODIFIED_DATE	System generated date of when the record was modified	DATETIME	0	NULL			
EMAIL	MODIFIED_BY	User ID of who modified the record	VARCHAR	50	NULL			
HABITAT_RESULT	HABITAT_RESULT_ID	Database generated unique ID (record identifier)	INT	0	NOT NULL	PRIMARY KEY		

Table 2-3. Continued.

TABLE_NAME	COLUMN_NAME	DESCRIPTION	DATA TYPE	LENGTH	REQUIRED	PRIMARY KEY	FOREIGN KEY	REFERENCE TABLE
HABITAT_RESULT	SAMPLING_EVENT_ID	ID from the SAMPLING_EVENT table that identifies the sample	INT	0	NOT NULL		FOREIGN KEY	SAMPLING EVENT
HABITAT_RESULT	HABITAT_ID	ID from the LV_HABITAT table that identifies the habitat parameter	INT	0	NOT NULL		FOREIGN KEY	LV HABITAT
HABITAT_RESULT	RESULT_VALUE	The measure result or value for the habitat parameter	DECIMAL	0	NOT NULL			
HABITAT_RESULT	RESULT_VALUE_UNIT_TYPE_ID	ID from the LV_UNIT_TYPE table that identifies the measurement unit	INT	0	NOT NULL		FOREIGN KEY	LV UNIT TYPE
HABITAT_RESULT	COMMENTS	Free text comments on a habitat result	VARCHAR	-1	NULL			
HABITAT_RESULT	CREATED_DATE	System generated date of when the record was created	DATETIME	0	NOT NULL			
HABITAT_RESULT	CREATED_BY	User ID of who created the record	VARCHAR	50	NOT NULL			
HABITAT_RESULT	MODIFIED_DATE	System generated date of when the record was modified	DATETIME	0	NULL			
HABITAT_RESULT	MODIFIED_BY	User ID of who modified the record	VARCHAR	50	NULL			

Table 2-3. Continued.

TABLE_NAME	COLUMN_NAME	DESCRIPTION	DATA TYPE	LENGTH	REQUIRED	PRIMARY KEY	FOREIGN KEY	REFERENCE TABLE
LOCATION	LOCATION_ID	Database generated unique ID (record identifier)	INT	0	NOT NULL	PRIMARY KEY		
LOCATION	STUDY_ID	Unique ID from the study table to link the study to the study location	INT	0	NOT NULL		FOREIGN KEY	STUDY
LOCATION	NAME	Text name or title for the study	VARCHAR	150	NOT NULL			
LOCATION	DESCRIPTION	Description of the study	VARCHAR	-1	NULL			
LOCATION	LOCATION_TYPE_ID	ID from the LV_LOCATION_TYPE table that identifies the type of location	INT	0	NOT NULL		FOREIGN KEY	LV LOCATION TYPE
LOCATION	LATITUDE	The latitude in Decimal Degrees	DECIMAL	0	NOT NULL			
LOCATION	LONGITUDE	The longitude in Decimal Degrees	DECIMAL	0	NOT NULL			

Table 2-3. Continued.

TABLE_NAME	COLUMN_NAME	DESCRIPTION	DATA TYPE	LENGTH	REQUIRED	PRIMARY KEY	FOREIGN KEY	REFERENCE TABLE
LOCATION	GEO_LOCATION	Geographic location ID	GEOGRAPHY	-1	NOT NULL			
LOCATION	COUNTY_ID	ID from the LV_COUNTY table that identifies the county	INT	0	NULL		FOREIGN KEY	LV COUNTY
LOCATION	STATE_ID	ID from the LV_STATE table that identifies the state	INT	0	NULL		FOREIGN KEY	LV STATE
LOCATION	COUNTRY_ID	ID from the LV_COUNTRY table that identifies the country	INT	0	NULL		FOREIGN KEY	LV COUNTRY
LOCATION	ZIP_CODE	usa zip code	VARCHAR	5	NULL			
LOCATION	HUC	Hydrologic Unit Code for the study location	VARCHAR	12	NULL			
LOCATION	STREAM_NAME	Text name for the stream	VARCHAR	150	NULL			
LOCATION	NHDS_STREAM_SEGMENT	National hydrologic data set (NHDS) stream segment code for the study location	VARCHAR	150	NULL			
LOCATION	CREATED_DATE	System generated date of when the record was created	DATETIME	0	NOT NULL			
LOCATION	CREATED_BY	User ID of who created the record	VARCHAR	50	NOT NULL			

Table 2-3. Continued.

TABLE_NAME	COLUMN_NAME	DESCRIPTION	DATA TYPE	LENGTH	REQUIRED	PRIMARY KEY	FOREIGN KEY	REFERENCE TABLE
LOCATION	MODIFIED_DATE	System generated date of when the record was modified	DATETIME	0	NULL			
LOCATION	MODIFIED_BY	User ID of who modified the record	VARCHAR	50	NULL			
LV_ANALYSIS_METHOD	ANALYSIS_METHOD_ID	Database generated unique ID (record identifier)	INT	0	NOT NULL	PRIMARY KEY		
LV_ANALYSIS_METHOD	ANALYSIS_METHOD_CONTEXT_ID	ID from the LV_ANALYSIS_METHOD_CONTEXT table that identifies the analysis method context.	INT	0	NULL		FOREIGN KEY	LV ANALYSIS METHOD CONTEXT
LV_ANALYSIS_METHOD	NAME	The name of the analysis method.	VARCHAR	150	NOT NULL			
LV_ANALYSIS_METHOD	DESCRIPTION	The description of the analysis method.	VARCHAR	-1	NULL			
LV_ANALYSIS_METHOD	REVIEWED_YN	Has the analysis method been reviewed (Y or N)	BIT	0	NOT NULL			
LV_ANALYSIS_METHOD	CREATED_DATE	System generated date of when the record was created	DATETIME	0	NOT NULL			
LV_ANALYSIS_METHOD	CREATED_BY	User ID of who created the record	VARCHAR	50	NOT NULL			
LV_ANALYSIS_METHOD	MODIFIED_DATE	System generated date of when the record was modified	DATETIME	0	NULL			

Table 2-3. Continued.

TABLE_NAME	COLUMN_NAME	DESCRIPTION	DATA TYPE	LENGTH	REQUIRED	PRIMARY KEY	FOREIGN KEY	REFERENCE TABLE
LV_ANALYSIS_METHOD	MODIFIED_BY	User ID of who modified the record	VARCHAR	50	NULL			
LV_ANALYSIS_METHOD_CONTEXT	ANALYSIS_METHOD_CONTEXT_ID	Database generated unique ID (record identifier)	INT	0	NOT NULL	PRIMARY KEY	FOREIGN KEY	LV ANALYSIS METHOD CONTEXT
LV_ANALYSIS_METHOD_CONTEXT	NAME	The name of the method context.	VARCHAR	150	NOT NULL			
LV_ANALYSIS_METHOD_CONTEXT	DESCRIPTION	The description of the method context.	VARCHAR	-1	NULL			
LV_ANALYSIS_METHOD_CONTEXT	REVIEWED_YN	Has the method context been reviewed (Y or N)	BIT	0	NOT NULL			
LV_ANALYSIS_METHOD_CONTEXT	CREATED_BY	System generated date of when the record was created	VARCHAR	50	NOT NULL			
LV_ANALYSIS_METHOD_CONTEXT	CREATED_DATE	User ID of who created the record	DATETIME	0	NOT NULL			
LV_ANALYSIS_METHOD_CONTEXT	MODIFIED_BY	System generated date of when the record was modified	VARCHAR	50	NULL			
LV_ANALYSIS_METHOD_CONTEXT	MODIFIED_DATE	User ID of who modified the record	DATETIME	0	NULL			
LV_ATTACHMENT_TABLE	ATTACHMENT_TABLE_ID	Database generated unique ID (record identifier)	INT	0	NOT NULL	PRIMARY KEY		
LV_ATTACHMENT_TABLE	NAME	The name of the attachment.	VARCHAR	150	NOT NULL			

Table 2-3. Continued.

TABLE_NAME	COLUMN_NAME	DESCRIPTION	DATA TYPE	LENGTH	REQUIRED	PRIMARY KEY	FOREIGN KEY	REFERENCE TABLE
LV_ATTACHMENT_TABLE	DESCRIPTION	The description of the attachment.	VARCHAR	-1	NULL			
LV_ATTACHMENT_TABLE	REVIEWED_YN	Has the attachment been reviewed for accuracy and applicability (Y or N)	CHAR	1	NOT NULL			
LV_ATTACHMENT_TABLE	CREATED_BY	User ID of who created the record	VARCHAR	50	NOT NULL			
LV_ATTACHMENT_TABLE	CREATED_DATE	System generated date of when the record was modified	DATETIME	0	NOT NULL			
LV_ATTACHMENT_TABLE	MODIFIED_BY	User ID of who modified the record	VARCHAR	50	NULL			
LV_ATTACHMENT_TABLE	MODIFIED_DATE	Database generated unique ID (record identifier)	DATETIME	0	NULL			
LV_BIOLOGICAL_INTENT	BIOLOGICAL_INTENT_ID	Database generated unique ID (record identifier)	INT	0	NOT NULL	PRIMARY KEY		
LV_BIOLOGICAL_INTENT	NAME	Name for the biological intent.	VARCHAR	150	NOT NULL			
LV_BIOLOGICAL_INTENT	DESCRIPTION	Description of the biological intent.	VARCHAR	-1	NULL			
LV_BIOLOGICAL_INTENT	REVIEWED_YN	Has the biological intent been reviewed (Y or N)	BIT	0	NOT NULL			

Table 2-3. Continued.

TABLE_NAME	COLUMN_NAME	DESCRIPTION	DATA TYPE	LENGTH	REQUIRED	PRIMARY KEY	FOREIGN KEY	REFERENCE TABLE
LV_BIOLOGICAL_INTENT	CREATED_DATE	System generated date of when the record was created	DATETIME	0	NOT NULL			
LV_BIOLOGICAL_INTENT	CREATED_BY	User ID of who created the record	VARCHAR	50	NOT NULL			
LV_BIOLOGICAL_INTENT	MODIFIED_DATE	System generated date of when the record was modified	DATETIME	0	NULL			
LV_BIOLOGICAL_INTENT	MODIFIED_BY	User ID of who modified the record	VARCHAR	50	NULL			
LV_COMPOUND	COMPOUND_ID	Database generated unique ID (record identifier)	INT	0	NOT NULL	PRIMARY KEY		
LV_COMPOUND	NAME	Name of the Compound or Substance	VARCHAR	150	NOT NULL			
LV_COMPOUND	DESCRIPTION	Description of the compound	VARCHAR	-1	NULL			
LV_COMPOUND	CAS_NUMBER	CAS registry numbers are unique numerical identifiers for chemical compounds.	VARCHAR	50	NOT NULL			
LV_COMPOUND	ENDOCRINE_DISRUPTING_C OMPOUND_YN	Is this an Endocrine disrupting compound (Y or N)	BIT	0	NOT NULL			
LV_COMPOUND	ENDOCRINE_ACTIVITY_VALU E	Measure of endocrine activity	INT	0	NULL			
LV_COMPOUND	TOXICITY_PBT_PROFILER	Predicted toxicity from PBT Profiler	DECIMAL	0	NULL			

Table 2-3. Continued.

TABLE_NAME	COLUMN_NAME	DESCRIPTION	DATA TYPE	LENGTH	REQUIRED	PRIMARY KEY	FOREIGN KEY	REFERENCE TABLE
LV_COMPOUND	CHV_VALUE_FISH_ECOSAR	Predicted chronic toxicity value to fish from ECOSAR	DECIMAL	0	NULL			
LV_COMPOUND	CHV_VALUE_DAPHNID_ECOSAR	Predicted chronic toxicity value to daphnids from ECOSAR	DECIMAL	0	NULL			
LV_COMPOUND	CHV_VALUE_ALGAE_ECOSAR	Predicted chronic toxicity value to algae from ECOSAR	DECIMAL	0	NULL			
LV_COMPOUND	MINIMUM_TOXICITY_VALUE	Lowest predicted toxicity value	DECIMAL	0	NULL			
LV_COMPOUND	HALF_LIFE_WATER_BPT_PROFILER	Half-life in water	DECIMAL	0	NULL			
LV_COMPOUND	HALF_LIFE_SOIL_PBT_PROFILER	Half-life in soil	DECIMAL	0	NULL			
LV_COMPOUND	HALF_LIFE_SEDIMENT_PBT_PROFILER	Half-life in sediment	DECIMAL	0	NULL			
LV_COMPOUND	HALF_LIFE_AIR_PBT_PROFILER	Half-life in air	DECIMAL	0	NULL			
LV_COMPOUND	LOG_BIOCONCENTRATION_CONSTANT_EPI_SUITE	Log bioconcentration constant	DECIMAL	0	NULL			
LV_COMPOUND	LOG_O_W_PARTITION_COEFFICIENT_SRC_ECOSAR	Log octanol water partition coefficient	DECIMAL	0	NULL			
LV_COMPOUND	WATER_SOLUBILITY_SCR	Solubility in water	DECIMAL	0	NULL			

Table 2-3. Continued.

TABLE_NAME	COLUMN_NAME	DESCRIPTION	DATA TYPE	LENGTH	REQUIRED	PRIMARY KEY	FOREIGN KEY	REFERENCE TABLE
LV_COMPOUND	SOIL_ADSORPTION_EPI_SUITE	Soil adsorption factor	DECIMAL	0	NULL			
LV_COMPOUND	HENRY_CONSTANT_SRC	Henry's constant	DECIMAL	0	NULL			
LV_COMPOUND	PRODUCTION_VOLUME_IN_US	Production volume in the United States	DECIMAL	0	NULL			
LV_COMPOUND	BIO_CONVERSION_FACTOR	Optional bioconversion factor	DECIMAL	0	NULL			
LV_COMPOUND	REVIEWED_YN	Has the compound been reviewed for accuracy and applicability (Y or N)	BIT	0	NOT NULL			
LV_COMPOUND	CREATED_DATE	System generated date of when the record was created	DATETIME	0	NOT NULL			
LV_COMPOUND	CREATED_BY	User ID of who created the record	VARCHAR	50	NOT NULL			
LV_COMPOUND	MODIFIED_DATE	System generated date of when the record was modified	DATETIME	0	NULL			
LV_COMPOUND	MODIFIED_BY	User ID of who modified the record	VARCHAR	50	NULL			
LV_COUNTRY	COUNTY_ID	Database generated unique ID (record identifier)	INT	0	NOT NULL	PRIMARY KEY		
LV_COUNTRY	NAME	Name of the country that appears in the data record	VARCHAR	150	NOT NULL			

Table 2-3. Continued.

TABLE_NAME	COLUMN_NAME	DESCRIPTION	DATA TYPE	LENGTH	REQUIRED	PRIMARY KEY	FOREIGN KEY	REFERENCE TABLE
LV_COUNTRY	DESCRIPTION	Description for the country	VARCHAR	-1	NULL			
LV_COUNTRY	REVIEWED_YN	Has country information been reviewed for accuracy (Y or N)	BIT	0	NOT NULL			
LV_COUNTRY	CREATED_DATE	System generated date of when the record was created	DATETIME	0	NOT NULL			
LV_COUNTRY	CREATED_BY	User ID of who created the record	VARCHAR	50	NOT NULL			
LV_COUNTRY	MODIFIED_DATE	System generated date of when the record was modified	DATETIME	0	NULL			
LV_COUNTRY	MODIFIED_BY	User ID of who modified the record	VARCHAR	50	NULL			
LV_COUNTY	COUNTY_ID	Database generated unique ID (record identifier)	INT	0	NOT NULL	PRIMARY KEY		
LV_COUNTY	NAME	The name of the county that appears in the data record	VARCHAR	150	NOT NULL			
LV_COUNTY	DESCRIPTION	A description of the county	VARCHAR	-1	NULL			

Table 2-3. Continued.

TABLE_NAME	COLUMN_NAME	DESCRIPTION	DATA TYPE	LENGTH	REQUIRED	PRIMARY KEY	FOREIGN KEY	REFERENCE TABLE
LV_COUNTY	STATE_ID	ID from the LV_STATE table that identifies the state associated with the county.	INT	0	NOT NULL		FOREIGN KEY	LV STATE
LV_COUNTY	REVIEWED_YN	Has county information been reviewed for accuracy (Y or N)	BIT	0	NOT NULL			
LV_COUNTY	CREATED_DATE	System generated date of when the record was created	DATETIME	0	NOT NULL			
LV_COUNTY	CREATED_BY	User ID of who created the record	VARCHAR	50	NOT NULL			
LV_COUNTY	MODIFIED_DATE	System generated date of when the record was modified	DATETIME	0	NULL			
LV_COUNTY	MODIFIED_BY	User ID of who modified the record	VARCHAR	50	NULL			
LV_DATA_QUALIFIER	DATA_QUALIFIER_ID	Database generated unique ID (record identifier)	INT	0	NOT NULL	PRIMARY KEY		
LV_DATA_QUALIFIER	NAME	Name of the data qualifier that appears in the data record.	VARCHAR	150	NOT NULL			
LV_DATA_QUALIFIER	DESCRIPTION	Description of the data qualifier.	VARCHAR	-1	NULL			

Table 2-3. Continued.

TABLE_NAME	COLUMN_NAME	DESCRIPTION	DATA TYPE	LENGTH	REQUIRED	PRIMARY KEY	FOREIGN KEY	REFERENCE TABLE
LV_DATA_QUALIFIER	REVIEWED_YN	Has the data qualifier information been reviewed for accuracy (Y or N)	BIT	0	NOT NULL			
LV_DATA_QUALIFIER	CREATED_DATE	System generated date of when the record was created	DATETIME	0	NOT NULL			
LV_DATA_QUALIFIER	CREATED_BY	User ID of who created the record	VARCHAR	50	NOT NULL			
LV_DATA_QUALIFIER	MODIFIED_DATE	System generated date of when the record was modified	DATETIME	0	NULL			
LV_DATA_QUALIFIER	MODIFIED_BY	User ID of who modified the record	VARCHAR	50	NULL			
LV_DETECTION_LIMIT_TYPE	DETECTION_LIMIT_TYPE_ID	Database generated unique ID (record identifier)	INT	0	NOT NULL	PRIMARY KEY		
LV_DETECTION_LIMIT_TYPE	NAME	The name of the detection limit that appears in the data record	VARCHAR	150	NOT NULL			
LV_DETECTION_LIMIT_TYPE	DESCRIPTION	A text description of the detection limit type	VARCHAR	-1	NULL			

Table 2-3. Continued.

TABLE_NAME	COLUMN_NAME	DESCRIPTION	DATA TYPE	LENGTH	REQUIRED	PRIMARY KEY	FOREIGN KEY	REFERENCE TABLE
LV_DETECTION_LIMIT_TYPE	REVIEWED_YN	Has detection limit type been reviewed for accuracy and applicability (Y or N)	BIT	0	NOT NULL			
LV_DETECTION_LIMIT_TYPE	CREATED_DATE	System generated date of when the record was created	DATETIME	0	NOT NULL			
LV_DETECTION_LIMIT_TYPE	CREATED_BY	User ID of who created the record	VARCHAR	50	NOT NULL			
LV_DETECTION_LIMIT_TYPE	MODIFIED_DATE	System generated date of when the record was modified	DATETIME	0	NULL			
LV_DETECTION_LIMIT_TYPE	MODIFIED_BY	User ID of who modified the record	VARCHAR	50	NULL			
LV_EVENT_TYPE	EVENT_TYPE_ID	Database generated unique ID (record identifier)	INT	0	NOT NULL	PRIMARY KEY		
LV_EVENT_TYPE	NAME	The name of the event type that appears in the data record	VARCHAR	150	NOT NULL			
LV_EVENT_TYPE	DESCRIPTION	A text description of the event type	VARCHAR	-1	NULL			
LV_EVENT_TYPE	REVIEWED_YN	Has event type been reviewed for accuracy and applicability (Y or N)	BIT	0	NOT NULL			

Table 2-3. Continued.

TABLE_NAME	COLUMN_NAME	DESCRIPTION	DATA TYPE	LENGTH	REQUIRED	PRIMARY KEY	FOREIGN KEY	REFERENCE TABLE
LV_EVENT_TYPE	CREATED_DATE	System generated date of when the record was created	DATETIME	0	NOT NULL			
LV_EVENT_TYPE	CREATED_BY	User ID of who created the record	VARCHAR	50	NOT NULL			
LV_EVENT_TYPE	MODIFIED_DATE	System generated date of when the record was modified	DATETIME	0	NULL			
LV_EVENT_TYPE	MODIFIED_BY	User ID of who modified the record	VARCHAR	50	NULL			
LV_HABIT	HABIT_ID	Database generated unique ID (record identifier)	INT	0	NOT NULL	PRIMARY KEY		
LV_HABIT	NAME	The name of the habit type that appears in the data record	VARCHAR	150	NOT NULL			
LV_HABIT	DESCRIPTION	A text description of the taxon habit	VARCHAR	-1	NULL			

Table 2-3. Continued.

TABLE_NAME	COLUMN_NAME	DESCRIPTION	DATA TYPE	LENGTH	REQUIRED	PRIMARY KEY	FOREIGN KEY	REFERENCE TABLE
LV_HABIT	REVIEWED_YN	Has habit been reviewed for accuracy and applicability (Y or N)	BIT	0	NOT NULL			
LV_HABIT	CREATED_DATE	System generated date of when the record was created	DATETIME	0	NOT NULL			
LV_HABIT	CREATED_BY	User ID of who created the record	VARCHAR	50	NOT NULL			
LV_HABIT	MODIFIED_DATE	System generated date of when the record was modified	DATETIME	0	NULL			
LV_HABIT	MODIFIED_BY	User ID of who modified the record	VARCHAR	50	NULL			
LV_HABITAT	HABITAT_ID	Database generated unique ID (record identifier)	INT	0	NOT NULL	PRIMARY KEY		
LV_HABITAT	NAME	The name of the habitat parameter that appears in the data record	VARCHAR	150	NOT NULL			
LV_HABITAT	DESCRIPTION	A text description of the habitat parameter	VARCHAR	-1	NULL			

Table 2-3. Continued.

TABLE_NAME	COLUMN_NAME	DESCRIPTION	DATA TYPE	LENGTH	REQUIRED	PRIMARY KEY	FOREIGN KEY	REFERENCE TABLE
LV_HABITAT	REVIEWED_YN	Has the habitat parameter been reviewed for accuracy and applicability (Y or N)	BIT	0	NOT NULL			
LV_HABITAT	CREATED_DATE	System generated date of when the record was created.	DATETIME	0	NOT NULL			
LV_HABITAT	CREATED_BY	User ID of who created the record.	VARCHAR	50	NOT NULL			
LV_HABITAT	MODIFIED_DATE	System generated date of when the record was modified.	DATETIME	0	NULL			
LV_HABITAT	MODIFIED_BY	User ID of who modified the record.	VARCHAR	50	NULL			
LV_LIFE_STAGE	LIFE_STAGE_ID	Database generated unique ID (record identifier).	INT	0	NOT NULL	PRIMARY KEY		
LV_LIFE_STAGE	NAME	The name of the life stage that appears in the data record.	VARCHAR	150	NOT NULL			
LV_LIFE_STAGE	DESCRIPTION	A text description of the habitat parameter.	VARCHAR	-1	NULL			

Table 2-3. Continued.

TABLE_NAME	COLUMN_NAME	DESCRIPTION	DATA TYPE	LENGTH	REQUIRED	PRIMARY KEY	FOREIGN KEY	REFERENCE TABLE
LV_LIFE_STAGE	REVIEWED_YN	Has the life stage designation been reviewed for accuracy and applicability (Y or N)	BIT	0	NOT NULL			
LV_LIFE_STAGE	CREATED_DATE	System generated date of when the record was created	DATETIME	0	NOT NULL			
LV_LIFE_STAGE	CREATED_BY	User ID of who created the record	VARCHAR	50	NOT NULL			
LV_LIFE_STAGE	MODIFIED_DATE	System generated date of when the record was modified	DATETIME	0	NULL			
LV_LIFE_STAGE	MODIFIED_BY	User ID of who modified the record	VARCHAR	50	NULL			
LV_LOCATION_TYPE	LOCATION_TYPE_ID	Database generated unique ID (record identifier)	INT	0	NOT NULL	PRIMARY KEY		
LV_LOCATION_TYPE	NAME	The name of the location type that appears in the data record	VARCHAR	150	NOT NULL			
LV_LOCATION_TYPE	DESCRIPTION	A text description of the location type.	VARCHAR	-1	NULL			
LV_LOCATION_TYPE	REVIEWED_YN	Has the location type been reviewed for accuracy and applicability (Y or N)	BIT	0	NOT NULL			

Table 2-3. Continued.

TABLE_NAME	COLUMN_NAME	DESCRIPTION	DATA TYPE	LENGTH	REQUIRED	PRIMARY KEY	FOREIGN KEY	REFERENCE TABLE
LV_LOCATION_TYPE	CREATED_DATE	System generated date of when the record was created	DATETIME	0	NOT NULL			
LV_LOCATION_TYPE	CREATED_BY	User ID of who created the record	VARCHAR	50	NOT NULL			
LV_LOCATION_TYPE	MODIFIED_DATE	System generated date of when the record was modified	DATETIME	0	NULL			
LV_LOCATION_TYPE	MODIFIED_BY	User ID of who modified the record	VARCHAR	50	NULL			
LV_MEASUREMENT_TYPE	MEASUREMENT_TYPE_ID	Database generated unique ID (record identifier)	INT	0	NOT NULL	PRIMARY KEY		
LV_MEASUREMENT_TYPE	NAME	The name of the measurement type that appears in the data record	VARCHAR	150	NOT NULL			
LV_MEASUREMENT_TYPE	DESCRIPTION	A text description of the measurement type.	VARCHAR	-1	NULL			
LV_MEASUREMENT_TYPE	REVIEWED_YN	Has the measurement type been reviewed for accuracy and applicability (Y or N)	BIT	0	NOT NULL			
LV_MEASUREMENT_TYPE	CREATED_DATE	System generated date of when the record was created	DATETIME	0	NOT NULL			
LV_MEASUREMENT_TYPE	CREATED_BY	User ID of who created the record	VARCHAR	50	NOT NULL			

Table 2-3. Continued.

TABLE_NAME	COLUMN_NAME	DESCRIPTION	DATA TYPE	LENGTH	REQUIRED	PRIMARY KEY	FOREIGN KEY	REFERENCE TABLE
LV_MEASUREMENT_TYPE	MODIFIED_DATE	System generated date of when the record was modified	DATETIME	0	NULL			
LV_MEASUREMENT_TYPE	MODIFIED_BY	User ID of who modified the record	VARCHAR	50	NULL			
LV_MEDIA_SUBDIVISION _TYPE	MEDIA_SUBDIVISION_TYPE_I D	Database generated unique ID (record identifier)	INT	0	NOT NULL	PRIMARY KEY		
LV_MEDIA_SUBDIVISION _TYPE	MEDIA_TYPE_ID	ID from the LV_MEDIA table that identifies the type of media.	INT	0	NOT NULL		FOREIGN KEY	LV MEDIA TYPE
LV_MEDIA_SUBDIVISION _TYPE	NAME	Name of the media at the subdivision level.	VARCHAR	150	NOT NULL			
LV_MEDIA_SUBDIVISION _TYPE	DESCRIPTION	A text description of the media at the subdivision level.	VARCHAR	-1	NULL			
LV_MEDIA_SUBDIVISION _TYPE	REVIEWED_YN	Has the media subdivision been reviewed for accuracy and applicability (Y or N)	BIT	0	NOT NULL			
LV_MEDIA_SUBDIVISION _TYPE	CREATED_DATE	System generated date of when the record was created	DATETIME	0	NOT NULL			
LV_MEDIA_SUBDIVISION _TYPE	CREATED_BY	User ID of who created the record	VARCHAR	50	NOT NULL			

Table 2-3. Continued.

TABLE_NAME	COLUMN_NAME	DESCRIPTION	DATA TYPE	LENGTH	REQUIRED	PRIMARY KEY	FOREIGN KEY	REFERENCE TABLE
LV_MEDIA_SUBDIVISION _TYPE	MODIFIED_DATE	System generated date of when the record was modified	DATETIME	0	NULL			
LV_MEDIA_SUBDIVISION _TYPE	MODIFIED_BY	User ID of who modified the record	VARCHAR	50	NULL			
LV_MEDIA_TYPE	MEDIA_TYPE_ID	Database generated unique ID (record identifier)	INT	0	NOT NULL	PRIMARY KEY		
LV_MEDIA_TYPE	NAME	The name of the media type that appears in the data record	VARCHAR	150	NULL			
LV_MEDIA_TYPE	DESCRIPTION	A text description of the media type.	VARCHAR	-1	NOT NULL			
LV_MEDIA_TYPE	REVIEWED_YN	Has the media been reviewed for accuracy and applicability (Y or N)	BIT	0	NOT NULL			
LV_MEDIA_TYPE	CREATED_DATE	System generated date of when the record was created	DATETIME	0	NOT NULL			
LV_MEDIA_TYPE	CREATED_BY	User ID of who created the record	VARCHAR	50	NOT NULL			

LV_MEDIA_TYPE	MODIFIED_DATE	System generated date of when the record was modified	DATETIME	0	NULL			
LV_MEDIA_TYPE	MODIFIED_BY	User ID of who modified the record	VARCHAR	50	NULL			
LV_SAMPLE_FRACTION	SAMPLE_FRACTION_ID	Database generated unique ID (record identifier)	INT	0	NOT NULL	PRIMARY KEY		
LV_SAMPLE_FRACTION	NAME	The name of the sample fraction that appears in the data record	VARCHAR	150	NOT NULL			
LV_SAMPLE_FRACTION	DESCRIPTION	A text description of the sample fraction	VARCHAR	-1	NULL			

Table 2-3. Continued.

TABLE_NAME	COLUMN_NAME	DESCRIPTION	DATA TYPE	LENGTH	REQUIRED	PRIMARY KEY	FOREIGN KEY	REFERENCE TABLE
LV_SAMPLE_FRACTION	REVIEWED_YN	Has the sample fraction been reviewed for accuracy and applicability (Y or N)	BIT	0	NOT NULL			
LV_SAMPLE_FRACTION	CREATED_DATE	System generated date of when the record was created	DATETIME	0	NOT NULL			
LV_SAMPLE_FRACTION	CREATED_BY	User ID of who created the record	VARCHAR	50	NOT NULL			
LV_SAMPLE_FRACTION	MODIFIED_DATE	System generated date of when the record was modified	DATETIME	0	NULL			
LV_SAMPLE_FRACTION	MODIFIED_BY	User ID of who modified the record	VARCHAR	50	NULL			
LV_SAMPLE_TYPE	SAMPLE_TYPE_ID	Database generated unique ID (record identifier)	INT	0	NOT NULL	PRIMARY KEY		
LV_SAMPLE_TYPE	NAME	The name of the sample type that appears in the data record	VARCHAR	150	NOT NULL			
LV_SAMPLE_TYPE	DESCRIPTION	A text description of the sample fraction	VARCHAR	-1	NULL			
LV_SAMPLE_TYPE	REVIEWED_YN	Has the sample type been reviewed for accuracy and applicability (Y or N)	BIT	0	NOT NULL			

Table 2-3. Continued.

TABLE_NAME	COLUMN_NAME	DESCRIPTION	DATA TYPE	LENGTH	REQUIRED	PRIMARY KEY	FOREIGN KEY	REFERENCE TABLE
LV_SAMPLE_TYPE	CREATED_DATE	System generated date of when the record was created	DATETIME	0	NOT NULL			
LV_SAMPLE_TYPE	CREATED_BY	User ID of who created the record	VARCHAR	50	NOT NULL			
LV_SAMPLE_TYPE	MODIFIED_DATE	System generated date of when the record was modified	DATETIME	0	NULL			
LV_SAMPLE_TYPE	MODIFIED_BY	User ID of who modified the record	VARCHAR	50	NULL			
LV_STATE	STATE_ID	Database generated unique ID (record identifier)	INT	0	NOT NULL	PRIMARY KEY		
LV_STATE	COUNTRY_ID	An id linking state to country for the LV_COUNTRY table	INT	0	NOT NULL		FOREIGN KEY	LV COUNTRY
LV_STATE	NAME	The name of the state that appears in the data record	VARCHAR	150	NOT NULL			
LV_STATE	CODE	state abbreviation	VARCHAR	8	NULL			
LV_STATE	DESCRIPTION	A text description of the state or state name	VARCHAR	-1	NULL			
LV_STATE	REVIEWED_YN	Has the state been reviewed for accuracy and applicability (Y or N)	BIT	0	NOT NULL			

Table 2-3. Continued.

TABLE_NAME	COLUMN_NAME	DESCRIPTION	DATA TYPE	LENGTH	REQUIRED	PRIMARY KEY	FOREIGN KEY	REFERENCE TABLE
LV_STATE	CREATED_DATE	System generated date of when the record was created.	DATETIME	0	NOT NULL			
LV_STATE	CREATED_BY	User ID of who created the record.	VARCHAR	50	NOT NULL			
LV_STATE	MODIFIED_DATE	System generated date of when the record was modified.	DATETIME	0	NULL			
LV_STATE	MODIFIED_BY	User ID of who modified the record.	VARCHAR	50	NULL			
LV_STUDY_STATUS	STUDY_STATUS_ID	Database generated unique ID (record identifier).	INT	0	NOT NULL	PRIMARY KEY		
LV_STUDY_STATUS	NAME	Text name or description of the study status.	VARCHAR	150	NOT NULL			
LV_STUDY_STATUS	DESCRIPTION	Text describing the meaning of the study status.	VARCHAR	-1	NULL			
LV_STUDY_STATUS	REVIEWED_YN	Has the study status been reviewed for accuracy (Y or N)	BIT	0	NOT NULL			
LV_STUDY_STATUS	CREATED_DATE	System generated date of when the record was created.	DATETIME	0	NOT NULL			
LV_STUDY_STATUS	CREATED_BY	User ID of who created the record.	VARCHAR	50	NOT NULL			

Table 2-3. Continued.

TABLE_NAME	COLUMN_NAME	DESCRIPTION	DATA TYPE	LENGTH	REQUIRED	PRIMARY KEY	FOREIGN KEY	REFERENCE TABLE
LV_STUDY_STATUS	MODIFIED_DATE	System generated date of when the record was modified.	DATETIME	0	NULL			
LV_STUDY_STATUS	MODIFIED_BY	User ID of who modified the record.	VARCHAR	50	NULL			
LV_TAXONOMY	TAXONOMY_ID	Database generated unique ID (record identifier)	INT	0	NOT NULL	PRIMARY KEY		
LV_TAXONOMY	TAXONOMY_PARENT_ID	ID from the LV_TAXONOMY table that identifies the parent taxon.	INT	0	NULL		FOREIGN KEY	LV TAXONOMY
LV_TAXONOMY	TAXONOMY_RANK_ID	ID from the LV_TAXONOMY_RANK table that identifies the rank (hierarchy of the taxa).	INT	0	NOT NULL		FOREIGN KEY	LV TAXONOMY RANK
LV_TAXONOMY	TAXONOMY_TYPE_ID	ID from the LV_TAXONOMY_TYPE table that identifies the type of taxon.	INT	0	NULL		FOREIGN KEY	LV TAXONOMY TYPE
LV_TAXONOMY	NAME	Name of the Taxon	VARCHAR	150	NOT NULL			
LV_TAXONOMY	DESCRIPTION	Description of the Taxon	VARCHAR	-1	NULL			

Table 2-3. Continued.

TABLE_NAME	COLUMN_NAME	DESCRIPTION	DATA TYPE	LENGTH	REQUIRED	PRIMARY KEY	FOREIGN KEY	REFERENCE TABLE
LV_TAXONOMY	REVIEWED_YN	Has the taxon name been reviewed for accuracy (Y or N)	BIT	0	NOT NULL			
LV_TAXONOMY	CREATED_DATE	System generated date of when the record was created	DATETIME	0	NOT NULL			
LV_TAXONOMY	CREATED_BY	User ID of who created the record	VARCHAR	50	NOT NULL			
LV_TAXONOMY	MODIFIED_DATE	System generated date of when the record was modified	DATETIME	0	NULL			
LV_TAXONOMY	MODIFIED_BY	User ID of who modified the record	VARCHAR	50	NULL			
LV_TAXONOMY_RANK	TAXONOMY_RANK_ID	Database generated unique ID (record identifier)	INT	0	NOT NULL	PRIMARY KEY		
LV_TAXONOMY_RANK	NAME	The name of the taxon rank that appears in the data record	VARCHAR	150	NOT NULL			
LV_TAXONOMY_RANK	DESCRIPTION	A text description of the taxon rank	VARCHAR	-1	NULL			
LV_TAXONOMY_RANK	REVIEWED_YN	Has the taxon name been reviewed for accuracy (Y or N)	BIT	0	NOT NULL			
LV_TAXONOMY_RANK	CREATED_DATE	System generated date of when the record was created	DATETIME	0	NOT NULL			

Table 2-3. Continued.

TABLE_NAME	COLUMN_NAME	DESCRIPTION	DATA TYPE	LENGTH	REQUIRED	PRIMARY KEY	FOREIGN KEY	REFERENCE TABLE
LV_TAXONOMY_RANK	CREATED_BY	User ID of who created the record	VARCHAR	50	NOT NULL			
LV_TAXONOMY_RANK	MODIFIED_DATE	System generated date of when the record was modified	DATETIME	0	NULL			
LV_TAXONOMY_RANK	MODIFIED_BY	User ID of who modified the record	VARCHAR	50	NULL			
LV_TAXONOMY_TYPE	TAXONOMY_TYPE_ID	Database generated unique ID (record identifier)	INT	0	NOT NULL	PRIMARY KEY		
LV_TAXONOMY_TYPE	NAME	The name of the type of taxon that appears in the data record	VARCHAR	150	NOT NULL			
LV_TAXONOMY_TYPE	DESCRIPTION	A text description of the taxon type.	VARCHAR	-1	NULL			
LV_TAXONOMY_TYPE	REVIEWED_YN	Has the taxonomy type been reviewed for accuracy and applicability (Y or N)	BIT	0	NOT NULL			
LV_TAXONOMY_TYPE	CREATED_DATE	System generated date of when the record was created	DATETIME	0	NOT NULL			
LV_TAXONOMY_TYPE	CREATED_BY	User ID of who created the record	VARCHAR	50	NOT NULL			
LV_TAXONOMY_TYPE	MODIFIED_DATE	System generated date of when the record was modified	DATETIME	0	NULL			

Table 2-3. Continued.

TABLE_NAME	COLUMN_NAME	DESCRIPTION	DATA TYPE	LENGTH	REQUIRED	PRIMARY KEY	FOREIGN KEY	REFERENCE TABLE
LV_TAXONOMY_TYPE	MODIFIED_BY	User ID of who modified the record	VARCHAR	50	NULL			
LV_TISSUE_ANATOMY	TISSUE_ANATOMY_ID	Database generated unique ID (record identifier)	INT	0	NOT NULL	PRIMARY KEY		
LV_TISSUE_ANATOMY	NAME	The name of the tissue anatomy that appears in the data record or result.	VARCHAR	150	NOT NULL			
LV_TISSUE_ANATOMY	DESCRIPTION	A text description of the tissue anatomy	VARCHAR	-1	NULL			
LV_TISSUE_ANATOMY	REVIEWED_YN	Has the tissue taxonomy been reviewed for accuracy and applicability (Y or N)	BIT	0	NOT NULL			
LV_TISSUE_ANATOMY	CREATED_DATE	System generated date of when the record was created	DATETIME	0	NOT NULL			
LV_TISSUE_ANATOMY	CREATED_BY	User ID of who created the record	VARCHAR	50	NOT NULL			
LV_TISSUE_ANATOMY	MODIFIED_DATE	System generated date of when the record was modified	DATETIME	0	NULL			
LV_TISSUE_ANATOMY	MODIFIED_BY	User ID of who modified the record	VARCHAR	50	NULL			

Table 2-3. Continued.

TABLE_NAME	COLUMN_NAME	DESCRIPTION	DATA TYPE	LENGTH	REQUIRED	PRIMARY KEY	FOREIGN KEY	REFERENCE TABLE
LV_TRACE_ORGANIC_TYPE	TRACE_ORGANIC_TYPE_ID	System generated unique id for each individual trace organic type.	INT	0	NOT NULL	PRIMARY KEY		
LV_TRACE_ORGANIC_TYPE	NAME	The name of the trace organic type that appears in the data record or result.	VARCHAR	150	NOT NULL			
LV_TRACE_ORGANIC_TYPE	DESCRIPTION	A text description of the type of trace organic.	VARCHAR	-1	NULL			
LV_TRACE_ORGANIC_TYPE	REVIEWED_YN	Has the trace organic type been reviewed for accuracy and applicability (Y or N)	BIT	0	NOT NULL			
LV_TRACE_ORGANIC_TYPE	CREATED_DATE	System generated date of when the record was created	DATETIME	0	NOT NULL			
LV_TRACE_ORGANIC_TYPE	CREATED_BY	User ID of who created the record	VARCHAR	50	NOT NULL			
LV_TRACE_ORGANIC_TYPE	MODIFIED_DATE	System generated date of when the record was modified	DATETIME	0	NULL			
LV_TRACE_ORGANIC_TYPE	MODIFIED_BY	User ID of who modified the record	VARCHAR	50	NULL			

Table 2-3. Continued.

TABLE_NAME	COLUMN_NAME	DESCRIPTION	DATA TYPE	LENGTH	REQUIRED	PRIMARY KEY	FOREIGN KEY	REFERENCE TABLE
LV_TROPHIC_LEVEL	TROPHIC_LEVEL_ID	Database generated unique ID (record identifier)	INT	0	NOT NULL	PRIMARY KEY		
LV_TROPHIC_LEVEL	NAME	The name of the trophic level that appears in the data record or result.	VARCHAR	150	NOT NULL			
LV_TROPHIC_LEVEL	DESCRIPTION	A text description of the trophic level.	VARCHAR	-1	NULL			
LV_TROPHIC_LEVEL	REVIEWED_YN	Has the unit conversion been reviewed for accuracy and applicability (Y or N)	BIT	0	NOT NULL			
LV_TROPHIC_LEVEL	CREATED_DATE	System generated date of when the record was created	DATETIME	0	NOT NULL			
LV_TROPHIC_LEVEL	CREATED_BY	User ID of who created the record	VARCHAR	50	NOT NULL			
LV_TROPHIC_LEVEL	MODIFIED_DATE	System generated date of when the record was modified	DATETIME	0	NULL			
LV_TROPHIC_LEVEL	MODIFIED_BY	User ID of who modified the record	VARCHAR	50	NULL			
LV_UNIT_CONVERSION	UNIT_CONVERSION_ID	Database generated unique ID (record identifier)	INT	0	NOT NULL	PRIMARY KEY		

Table 2-3. Continued.

TABLE_NAME	COLUMN_NAME	DESCRIPTION	DATA TYPE	LENGTH	REQUIRED	PRIMARY KEY	FOREIGN KEY	REFERENCE TABLE
LV_UNIT_CONVERSION	NAME	The name of the unit conversion that appears in the data record or result.	VARCHAR	150	NULL			
LV_UNIT_CONVERSION	DESCRIPTION	A text description of the unit conversion.	VARCHAR	-1	NULL			
LV_UNIT_CONVERSION	UNIT_TYPE_1_ID	The before conversion unit type.	INT	0	NOT NULL			
LV_UNIT_CONVERSION	UNIT_TYPE_2_ID	The converted unit type.	INT	0	NOT NULL			
LV_UNIT_CONVERSION	CONVERSION_FACTOR	The factor needed to convert the original unit to the desired unit.	DECIMAL	0	NOT NULL			
LV_UNIT_CONVERSION	REVIEWED_YN	Has the trophic level been reviewed for accuracy and applicability (Y or N)	BIT	0	NOT NULL			
LV_UNIT_CONVERSION	CREATED_DATE	System generated date of when the record was created	DATETIME	0	NOT NULL			
LV_UNIT_CONVERSION	CREATED_BY	User ID of who created the record	VARCHAR	50	NOT NULL			
LV_UNIT_CONVERSION	MODIFIED_DATE	System generated date of when the record was modified	DATETIME	0	NULL			

Table 2-3. Continued.

TABLE_NAME	COLUMN_NAME	DESCRIPTION	DATA TYPE	LENGTH	REQUIRED	PRIMARY KEY	FOREIGN KEY	REFERENCE TABLE
LV_UNIT_CONVERSION	MODIFIED_BY	User ID of who modified the record	VARCHAR	50	NULL			
LV_UNIT_TYPE	UNIT_TYPE_ID	Database generated unique ID (record identifier)	INT	0	NOT NULL	PRIMARY KEY		
LV_UNIT_TYPE	NAME	The name of the unit type that appears in the record or result.	VARCHAR	150	NOT NULL			
LV_UNIT_TYPE	DESCRIPTION	A text description of the unit type.	VARCHAR	-1	NULL			
LV_UNIT_TYPE	REVIEWED_YN	Has the unit type been reviewed for accuracy applicability (Y or N)	BIT	0	NOT NULL			
LV_UNIT_TYPE	CREATED_DATE	System generated date of when the record was created	DATETIME	0	NOT NULL			
LV_UNIT_TYPE	CREATED_BY	User ID of who created the record	VARCHAR	50	NOT NULL			
LV_UNIT_TYPE	MODIFIED_DATE	System generated date of when the record was modified	DATETIME	0	NULL			

Table 2-3. Continued.

TABLE_NAME	COLUMN_NAME	DESCRIPTION	DATA TYPE	LENGTH	REQUIRED	PRIMARY KEY	FOREIGN KEY	REFERENCE TABLE
LV_UNIT_TYPE	MODIFIED_BY	User ID of who modified the record	VARCHAR	50	NULL			
LV USE AND OR SECONDARY TYPE	USE_AND_OR_SECONDARY_TYPE_ID	Database generated unique ID (record identifier)	INT	0	NOT NULL	PRIMARY KEY		
LV_USE_AND_OR_SECONDARY_TYPE	NAME	The name of the use or secondary type that appears in the record or result.	VARCHAR	150	NOT NULL			
LV_USE_AND_OR_SECONDARY_TYPE	DESCRIPTION	A text description of the use or secondary type	VARCHAR	-1	NULL			
LV_USE_AND_OR_SECONDARY_TYPE	REVIEWED_YN	Has the use or secondary type been reviewed for accuracy and applicability (Y or N)	BIT	0	NOT NULL			
LV_USE_AND_OR_SECONDARY_TYPE	CREATED_DATE	System generated date of when the record was created	DATETIME	0	NOT NULL			
LV_USE_AND_OR_SECONDARY_TYPE	CREATED_BY	User ID of who created the record	VARCHAR	50	NOT NULL			
LV_USE_AND_OR_SECONDARY_TYPE	MODIFIED_DATE	System generated date of when the record was modified	DATETIME	0	NULL			
LV_USE_AND_OR_SECONDARY_TYPE	MODIFIED_BY	User ID of who modified the record	VARCHAR	50	NULL			

Table 2-3. Continued.

TABLE_NAME	COLUMN_NAME	DESCRIPTION	DATA TYPE	LENGTH	REQUIRED	PRIMARY KEY	FOREIGN KEY	REFERENCE TABLE
LV_WEIGHT_BASIS	WEIGHT_BASIS_ID	Database generated unique ID (record identifier)	INT	0	NOT NULL	PRIMARY KEY		
LV_WEIGHT_BASIS	NAME	The name of the weight basis that appears in the record or result.	VARCHAR	150	NOT NULL			
LV_WEIGHT_BASIS	DESCRIPTION	A text description of the weight basis	VARCHAR	-1	NULL			
LV_WEIGHT_BASIS	REVIEWED_YN	Has the weight basis been reviewed for accuracy and applicability (Y or N)	BIT	0	NOT NULL			
LV_WEIGHT_BASIS	CREATED_DATE	System generated date of when the record was created	DATETIME	0	NOT NULL			
LV_WEIGHT_BASIS	CREATED_BY	User ID of who created the record	VARCHAR	50	NOT NULL			
LV_WEIGHT_BASIS	MODIFIED_DATE	System generated date of when the record was modified	DATETIME	0	NULL			
LV_WEIGHT_BASIS	MODIFIED_BY	User ID of who modified the record	VARCHAR	50	NULL			
SAMPLING_EVENT	SAMPLING_EVENT_ID	Database generated unique ID (record identifier)	INT	0	NOT NULL	PRIMARY KEY		

Table 2-3. Continued.

TABLE_NAME	COLUMN_NAME	DESCRIPTION	DATA TYPE	LENGTH	REQUIRED	PRIMARY KEY	FOREIGN KEY	REFERENCE TABLE
SAMPLING_EVENT	LOCATION_ID	ID from the LOCATION table that identifies the status of the study.	INT	0	NOT NULL		FOREIGN KEY	LOCATION
SAMPLING_EVENT	START_DATE	The start date for the sampling activity or event.	DATETIME	0	NOT NULL			
SAMPLING_EVENT	END_DATE	The end date for the sampling activity or event.	DATETIME	0	NULL			
SAMPLING_EVENT	START_TIME	The start time for the sampling activity or event.	DATETIME	0	NULL			
SAMPLING_EVENT	END_TIME	The end time for the sampling activity or event.	DATETIME	0	NULL			
SAMPLING_EVENT	COMMENTS	Free text comments on the sampling event.	VARCHAR	-1	NULL			
SAMPLING_EVENT	EVENT_TYPE_ID	ID from the LV_EVENT_TYPE table that identifies the type of sampling event.	INT	0	NULL		FOREIGN KEY	LV EVENT TYPE
SAMPLING_EVENT	SAMPLE_TYPE_ID	ID from the LV_SAMPLE_TYPE table that identifies the type of sampling event.	INT	0	NOT NULL		FOREIGN KEY	LV SAMPLE TYPE
SAMPLING_EVENT	CREATED_DATE	System generated date of when the record was created	DATETIME	0	NOT NULL			

Table 2-3. Continued.

TABLE_NAME	COLUMN_NAME	DESCRIPTION	DATA TYPE	LENGTH	REQUIRED	PRIMARY KEY	FOREIGN KEY	REFERENCE TABLE
SAMPLING_EVENT	CREATED_BY	User ID of who created the record	VARCHAR	50	NOT NULL			
SAMPLING_EVENT	MODIFIED_DATE	System generated date of when the record was modified	DATETIME	0	NULL			
SAMPLING_EVENT	MODIFIED_BY	User ID of who modified the record	VARCHAR	50	NULL			
STUDY	STUDY_ID	A database generated unique ID (record identifier).	INT	0	NOT NULL	PRIMARY KEY		
STUDY	PRIMARY_AUTHOR	The primary author of the study.	VARCHAR	150	NOT NULL			
STUDY	SECONDARY_AUTHOR	The secondary author of the study.	VARCHAR	150	NULL			
STUDY	JOURNAL	The name of the journal the study is published.	VARCHAR	250	NULL			
STUDY	TITLE	Title of the study or paper.	VARCHAR	250	NULL			
STUDY	YEAR	The year the study was published.	VARCHAR	30	NULL			
STUDY	VOLUME	The volume of the journal in which the study is published.	VARCHAR	30	NULL			

Table 2-3. Continued.

TABLE_NAME	COLUMN_NAME	DESCRIPTION	DATA TYPE	LENGTH	REQUIRED	PRIMARY KEY	FOREIGN KEY	REFERENCE TABLE
STUDY	PAGES	The page numbers for the paper or study in the journal article.	VARCHAR	30	NULL			
STUDY	URL	The URL where the study can be found on line.	VARCHAR	300	NULL			
STUDY	PUBLISHED_STUDY_YN	Was the paper published (Y or N)	BIT	0	NOT NULL			
STUDY	PRIMARY_STUDY_YN	Is this a primary study (Y or N)	BIT	0	NOT NULL			
STUDY	QA_QC_STUDY_YN	Has the study been QCed (Y or N)	BIT	0	NOT NULL			
STUDY	QUALITY_OF_COORDINATES_YN	Is the quality of the latitude and Longitude coordinates known	BIT	0	NOT NULL			
STUDY	STUDY_STATUS_ID	ID from the LV_STUDY_STATUS table that identifies the status of the study.	INT	0	NOT NULL		FOREIGN KEY	LV STUDY STATUS
STUDY	KEYWORDS	Keywords from the study or paper.	VARCHAR	-1	NOT NULL			
STUDY	ORGANIZATION_NAME	Name of the organization conducting or associated with the study.	VARCHAR	150	NOT NULL			

Table 2-3. Continued.

TABLE_NAME	COLUMN_NAME	DESCRIPTION	DATA TYPE	LENGTH	REQUIRED	PRIMARY KEY	FOREIGN KEY	REFERENCE TABLE
STUDY	ORGANIZATION_ADDRESS_LINE_1	Address line for the organization.	VARCHAR	100	NULL			
STUDY	ORGANIZATION_ADDRESS_LINE_2	Address line for the organization.	VARCHAR	100	NULL			
STUDY	ORGANIZATION_CITY	City address for the organization.	VARCHAR	50	NULL			
STUDY	ORGANIZATION_STATE_ID	State address for the organization.	INT	0	NULL		FOREIGN KEY	LV STATE
STUDY	ORGANIZATION_COUNTRY_ID	Country address for the organization.	INT	0	NULL		FOREIGN KEY	LV COUNTRY
STUDY	ORGANIZATION_PHONE	Contact phone number for the organization.	VARCHAR	20	NULL			
STUDY	ORGANIZATION_EMAIL	Email address for the organization.	VARCHAR	50	NULL			
STUDY	ORGANIZATION_ZIP_CODE	Address zip code for the organization.	VARCHAR	10	NULL			
STUDY	CREATED_DATE	System generated date of when the record was created.	DATETIME	0	NOT NULL			
STUDY	CREATED_BY	User ID of who created the record.	VARCHAR	50	NOT NULL			

Table 2-3. Continued.

TABLE_NAME	COLUMN_NAME	DESCRIPTION	DATA TYPE	LENGTH	REQUIRED	PRIMARY KEY	FOREIGN KEY	REFERENCE TABLE
STUDY	MODIFIED_DATE	*System generated date of* when the record was modified.	DATETIME	0	NULL			
STUDY	MODIFIED_BY	User ID of who modified the record.	VARCHAR	50	NULL			
STUDY_KEYWORD	STUDY_KEYWORD_ID	Database generated unique ID (record identifier).	INT	0	NOT NULL	PRIMARY KEY		
STUDY_KEYWORD	STUDY_ID	An id that links the study keywords to the study and study table.	INT	0	NOT NULL		FOREIGN KEY	STUDY
STUDY_KEYWORD	KEYWORD	Keywords from the linked study.	VARCHAR	50	NOT NULL			

GLOSSARY

Term	Project Definition	Source
Abiotic	Not living, non-biological, usually describing factors in an ecosystem such as atmospheric gases, inorganic salts, mineral soil particles, and water. This expression is also used to describe the chemical and physical factors, such as salinity and humidity, which influence organisms. An *abiotic environment* is one without any life.	http://www.answers.com/topic/abiotic
Acute to Chronic Ratio	The species mean acute toxicity divided by the chronic toxicity for the same species. The final acute to chronic ratio is the geometric mean of all acute to chronic ratios for species of animals in at least three different families.	*Canadian Environmental Quality Guidelines,* 1999
Adverse ecological effects	Changes that are considered undesirable because they alter valued structural or functional characteristics of ecosystems or their components. An evaluation of adversity may consider the type, intensity, and scale of the effect as well as the potential for recovery.	http://cfpub.epa.gov/ncea/CFM/recordisplay.cfm?deid=12460
Bioaccumulation	Process by which a compound is taken up by an aquatic organism, both from water and through food . This process may result in an increase in levels of toxic substances in an organism over time, due to continued exposure	http://www.epa.gov/waterscience/standards/handbook/glossary.html www.ec.gc.ca/water/en/info/gloss/e_gloss.htm
Biological assessment	Evaluation of the biological condition of a waterbody using biological surveys and other direct measurements of resident biota in surface waters.	EPA glossary http://www.epa.gov/waterscience/biocriteria/glossary.html#biologicalassessments
Bioconcentration	The process by which contaminants are directly taken up by organisms from the water in which they live.	*Canadian Environmental Quality Guidelines,* 1999
Biodegradable	Capable of being broken down by living organisms into inorganic compounds.	www.ec.gc.ca/water/en/info/gloss/e_gloss.htm
Biomagnification	The increase in tissue concentrations of accumulated chemicals from one trophic level to the next (i.e., organisms contain higher concentrations of the substance than their food sources).	*Canadian Environmental Quality Guidelines,* 1999
Biological monitoring	The use of living organisms in water quality surveillance to indicate compliance with water quality standards or effluent limits and to document water quality trends. Methods of biological monitoring may include, but are not limited to, toxicity testing (such as ambient toxicity testing or whole-effluent toxicity testing), biomarker analyses, and biological surveys. It is also known as biomonitoring.	http://www.epa.gov/waterscience/standards/handbook/glossary.html
Biomarkers	Biochemical or cellular indicators of exposure to contaminants (e.g., body burden, indicators of DNA damage, enzyme activity, and biochemical indicators of reproductive or biogenetic status). Also referred to as xenobiotically-induced variation in cellular or biochemical components or processes, structures or functions that is measurable in a biological system or sample	*Canadian Environmental Quality Guidelines,* 1999 Shugart et al. 1992
Biological survey	Collecting, processing, and analyzing a representative portion of the resident aquatic community to determine its structural and/or functional characteristics	http://www.epa.gov/waterscience/standards/handbook/glossary.html

Term	Project Definition	Source
Biota	The animal and plant life of a given region. Collectively, the plants, microorganisms, and animals of a certain area or region.	http://www.epa.gov/OCEPAterms/aterms.html www.ec.gc.ca/water/en/info/gloss/e_gloss.htm
Candidate cause	A hypothesized cause of an environmental impairment which is sufficiently credible to be analyzed.	USEPA 822-B-00-025 Stressor Identification Guidance Document, 2000
Causal analysis	a process in which data and other information are organized and evaluated using quantitative and logical techniques to determine the likely cause of an observed condition	USEPA 822-B-00-025 Stressor Identification Guidance Document, 2000
Cause	1. that which produces an effect (general definition). 2. a stressor or set of stressors that occur at an intensity, duration and frequency of exposure that results in a change in the ecological condition (specific definition for this research).	USEPA 822-B-00-025 Stressor Identification Guidance Document, 2000
Community (aquatic community)	An association of interacting populations of aquatic organisms in a given water body or habitat.	http://www.epa.gov/waterscience/standards/handbook/glossary.html
Conceptual model	A written description and visual representation of predicted relationships between ecological entities and the stressors to which they may be exposed	http://cfpub.epa.gov/ncea/CFM/recordisplay.cfm?deid=12460
Contaminant	Any unnatural biological, chemical, physical, or radiological substance or matter contained in water	www.water-research.net/glossary.htm
Cumulative environmental effects	The impact on the environment which results from the incremental impact of the action when added to other past, present, and reasonably foreseeable future actions.	US Council of Environmental Quality (CEQ) regulations in 40 CFR Part 1508.7
Diagnostic analysis	A type of causal analysis in which effects that are characteristic of a particular cause are used to determine whether that candidate cause may be responsible for impairment.	USEPA 822-B-00-025 Stressor Identification Guidance Document, 2000
Ecological Indicator	A characteristic of an ecosystem that is related to, or derived from, a measure of biotic or abiotic variable, that can provide quantitative information on ecological structure and function. An indicator can contribute to a measure of integrity and sustainability.	http://www.epa.gov/OCEPAterms/eterms.html
Ecological risk assessment	The application of a formal framework, analytical process, or model to estimate the effects of human actions(s) on a natural resource and to interpret the significance of those effects in light of the uncertainties identified in each component of the assessment process. Such analysis includes initial hazard identification, exposure and dose-response assessments, and risk characterization.	http://www.epa.gov/OCEPAterms/aterms.html
Ecosystem	**A** dynamic complex of plant, animal, and microorganism communities and the nonliving environment interacting as a functional unit.	www.env.gov.bc.ca/eirs/bdp/glossary.html
Ecosystem function	Ecosystem function includes inputs, outputs, cycling of materials and energy, and the interactions of organisms. Processes include nutrient cycling, primary production, and detrital processing	http://www.esa.org/pao/policyStatements/Papers/ReportOfSBEM.php

Term	Project Definition	Source
Ecosystem Structure	Attributes related to the instantaneous physical state of an ecosystem; examples include species population density, species richness or evenness, and standing crop biomass.	http://www.epa.gov/OCEPAterms/eterms.html
Effects (ecological effects)	Something that inevitably follows an antecedent (cause or agent). A biological effect is the biological result of exposure to a causal agent. ***Direct effects*** are the induction of an *effect* through a single cause-effect relationship. ***Indirect effects*** are changes in a resource that are due to a series of cause-effect relationships rather than to direct exposure to a contaminant or other stressor.	http://cfpub.epa.gov/caddis/info_sources.cfm?Section=30
Effluent	Wastewater--treated or untreated--that flows out of a treatment plant, sewer, or industrial outfall. Generally refers to wastes discharged into surface waters.	http://www.epa.gov/OCEPAterms/aterms.html
Endocrine disrupting compound (EDC)	Chemicals or substances that disrupt normal endocrine system functions in humans or wildlife by blocking or mimicking hormones.	http://www.chesapeakebay.net/glossary.aspx?menuitem=14875#I
Endpoint (assessment endpoint)	In ecological risk assessment, an explicit expression of the environmental value to be protected; includes both an ecological entity and specific attribute thereof. e.g., salmon are a valued ecological entity; reproduction and population maintenance--the attribute--form an assessment endpoint.)	http://www.epa.gov/OCEPAterms/aterms.html
Exposure	The amount of a pollutant present in a given environment or at the site of action within an organism that represents a potential health threat to living organisms; the co-occurrence or contact of a stressor and the resource that becomes impaired	http://www.epa.gov/OCEPAterms/aterms.html USEPA 822-B-00-025 Stressor Identification Guidance Document, 2000
Exposure–response pathway	The process by which an individual or population is exposed to contaminants and experiences an effect from this exposure.	http://www.epa.gov/waterscience/
Exposure–response relationship	A qualitative or quantitative (usually statistical) model of the relationship between an exposure metric (e.g., the concentration of a chemical or the abundance or an exotic species) to which a population or community of organisms is exposed and the frequency or magnitude of a biological response.	USEPA 822-B-00-025 Stressor Identification Guidance Document, 2000
Fecundity	The number of eggs produced per female during a spawning season.	http://www.chesapeakebay.net/glossary.aspx?menuitem=14875#I
Habitat	A place where the physical and biological elements of ecosystems provide a suitable environment including the food, cover, and space resources needed for plant and animal livelihood.	http://www.epa.gov/waterscience/biocriteria/glossary.html
Impairment	Detrimental effect on the biological integrity of a waterbody caused by an impact that prevents attainment of the designated use, which would include ecosystem structure and function.	http://www.epa.gov/waterscience/biocriteria/glossary.html#biologicalassessments
Indicators	Groups or types of biological resources (i.e., fish, invertebrates, periphyton, macrophytes) or endpoints that can be used to assess environmental condition.	http://www.epa.gov/bioiweb1/html/indicator.html

Term	Project Definition	Source
Intersex	An animal (usually a fish) that displays both male and female characteristics. Intersex in fish has been linked to exposure to hormone-disrupting compounds such as pharmaceuticals, personal care products, herbicides, pesticides and other agricultural products.	http://www.chesapeakebay.net/glossary.aspx?menuitem=14875#I
Mixing zone	An area where an effluent discharge undergoes initial dilution and is extended to cover the secondary mixing in the ambient water body. A mixing zone is an allocated impact zone where water quality criteria can be exceeded as long as acutely toxic conditions are prevented.	http://cfpub.epa.gov/npdes/glossary.cfm?program_id=0#N
Mode of Action	Also referred to as mechanism of action. The mechanisms by which a chemical enters and organism, interacts with target molecules, and causes deleterious effects.	Casarett & Doull's Toxicology. 5th Edition, 1996, C. Klaussen, ed.
National Pollutant Discharge Elimination System (NPDES)	A national program under Section 402 of the Clean Water Act for regulation of discharges of pollutants from point sources to waters of the United States. Discharges are illegal unless authorized by an NPDES permit.	http://cfpub.epa.gov/npdes/glossary.cfm?program_id=0#N
Nonpoint source	A diffuse source of pollution that cannot be attributed to a clearly identifiable, specific physical location or a defined discharge channel. This includes the nutrients that runoff the ground from any land use croplands, feedlots, lawns, parking lots, streets, forests, etc.	http://www.werf.org/livablecommunities/gloss.htm
Pathway	The physical course a chemical or pollutant takes from its source to the exposed organism.	http://www.epa.gov/OCEPAterms/aterms.html
Persistence	Refers to the length of time a compound stays in the environment, once introduced. A compound may persist for less than a second up to indefinitely.	http://www.epa.gov/OCEPAterms/aterms.html
Pollutant	Dredged spoil, solid waste, incinerator residue, sewage, garbage, sewage sludge, munitions, chemical wastes, biological materials, radioactive materials, heat, wrecked or discarded equipment, rock, sand, cellar dirt, and industrial, municipal, and agricultural wastes discharged into water.	http://www.epa.gov/owow/tmdl/glossary.html#pollutant
Point source	Water pollution sources that may be traced to a specific source, such as a sewer line or a discharge pipe of an industrial facility. Pollution coming from a single identifiable source such as discharge pipes from industry or sewer plants or other means of conveyance including ditches, channels, sewers, and containers.	http://www.werf.org/livablecommunities/gloss.htm
Population	Aggregate of individuals of a biological species that are geographically isolated from other members of the species and are actually or potentially interbreeding.	http://www.epa.gov/waterscience/biocriteria/glossary.html#biologicalassessments
Pseudo-persistence	Continual presence of a non-persistent chemical at a site due to constant discharge or input of the chemical from a source or sources	
Publicly Owned Treatment Works (POTWs)	A treatment works, as defined by Section 212 of the CWA, that is owned by the state or municipality. This definition includes any devices and systems used in the storage, treatment, recycling, and reclamation of municipal sewage or industrial wastes of a liquid nature. Privately-owned treatment works, Federally-owned treatment works, and other treatment plants not owned by municipalities are not considered POTWs.	http://cfpub.epa.gov/npdes/glossary.cfm?program_id=0#P

Term	Project Definition	Source
Receiving waters	A river, lake, ocean, stream or other watercourse into which pollutants, wastewater or treated effluent is discharged.	http://www.epa.gov/OCEPAterms/aterms.html
QSAR (quantitative structural activity relationship)	A method of estimating the unmeasured physical and toxicological properties of a chemical through chemical structure, functional groups, and similarities to known chemicals.	*Canadian Environmental Quality Guidelines*, 1999
Receptor	Organisms, populations, or ecosystems that are exposed to a contaminant or other stressor.	EPA/822/B-00/025 Stressor Identification Guidance Document, 2000
Resident species	A species that is present at a specified site for some portion of its life span.	ASTM E-1850; Standard Guide to Selection of Resident Species for Use in Toxicity Testing
Risk assessment	Qualitative and quantitative evaluation of the risk posed to human health and/or the environment by the actual or potential presence and/or use of specific pollutants. *Prospective risk assessment* : a risk assessment performed for a proposed future action or potential change in the environment *Retrospective risk assessment*: a risk assessment performed for hazards or impairments that began in the past and may have ongoing effects.	http://www.epa.gov/OCEPAterms/aterms.html G. Suter 1993. Ecological Risk Assessment. Lewis Publishers
Risk hypothesis	Theoretical relationships or linkages from source through stressor that might contribute to the likelihood of unacceptable harmful effects.	A. Raybould 2006 http://www.ebr-journal.org/index.php?option=article&access=standard&Itemid=129&url=/articles/ebr/pdf/2006/03/ebr0638.pdf
Stressor	Physical, chemical, or biological entities that can induce adverse effects on biota, ecosystems or human health.	http://www.epa.gov/OCEPAterms/aterms.html
Synergistic interaction	Interaction of discreet agents (chemicals) and/or conditions (temperature) that results in an effect that is greater than the sum of the effects of the individual agents and/or conditions.	www.merriam-webster.com
Threshold	The dose or exposure level below which a significant adverse effect is not expected for a given species or type of biota.	http://www.epa.gov/OCEPAterms/aterms.html
TOrC (trace organic chemicals)	Chemicals that are known or suspected to be released to aquatic environments but are not commonly regulated or monitored, and whose potential risk to human or ecological health is relatively unknown.	WERF project website
Toxicity	The degree to which a substance or mixture of substances can harm humans or animals. *Acute toxicity* involves short-term harmful effects in an organism such as mortality or immobility. *Chronic toxicity* is the ability of a substance or mixture of substances to cause harmful sublethal effects, usually upon repeated or continuous exposure.	http://www.epa.gov/OCEPAterms/aterms.html
Toxicity test	A procedure to determine the toxicity of a chemical or an effluent using living organisms. A toxicity test measures the degree of effect on exposed test organisms of a specific chemical or effluent.	http://cfpub.epa.gov/npdes/glossary.cfm?program_id=0#T

Term	Project Definition	Source
Vitellogenin	An egg yolk protein precursor that is produced by liver cells in response to estrogen or estrogen mimics and is used to mature oocytes. Vitellogenin is normally only found in female fish but it is also being used to assess whether male fish are exposed to estrogens in the environment.	Sumpter, J. and S. Jobling 1995. Vitellogenesis as a Biomarker for Estrogenic Contamination of the Aquatic Environment. Environ. Health Perspect. 103:173-178.
Wastewater	Water that carries wastes from homes, businesses, and industries; a mixture of water and dissolved or suspended solids	www.ec.gc.ca/water/en/info/gloss/e_gloss.htm
Watershed	Basins drained by or contributing water to a particular point, stream, river, lake or ocean	www.water-research.net/glossary.htm

REFERENCES

Alvarez, D., W. Cranor, S. Perkins, V. Schroeder, L. Iwanowicz, R. Clark, C. Guy, A. Pinkney, V. Blazer, and J. Mullican. 2009. Reproductive health of bass in the Potomac, USA, drainage: Part 2. Seasonal occurrence of persistent and emerging organic contaminants. *Environ. Toxicol. Chem.* 28:1084-1095.

Ankley, G.T., G.P. Daston, S.J. Degitz, N.D. Denslow, R.A. Hoke, S.W. Kennedy, A.L. Miracle, E.J. Perkins, J. Snape, D.E. Tillitt, C.R. Tyler, and D. Versteeg. 2006. Toxicogenomics in regulatory ecotoxicology. *Environ. Sci. Technol.* 40, 4055-4065.

Ankley, G., D. Bencic, M. Breen, T. Collette, R. Conolly, N. Denslow, S. Edwards, D. Ekman, N. Garcia-Reyero, K. Jensen, J. Lazorchak, D. Martinovic, D. Miller, E. Perkins, E. Orlando, D. Villeneuve, R. Wang, and K. Watanabe. 2009. Endocrine disrupting chemicals in fish: Developing exposure indicators and predictive models of effects based on mechanism of action. *Aquat. Toxicol.* 92:168-178.

Bay, S. 2010. Sources and effects of endocrine disruptors and other contaminants of emerging concern in the Southern California Bight Coastal Ecosystem. Southern California Coastal Water Research Project., Costa Mesa, CA. Available at: http://www.sccwrp.org/RresearchAreas/Contaminants/ContaminantsOfEmergingConcern/EffectsOnBiota/EmergingContaminantEffectsOnCoastalFish.aspx

Blazer, V., L. Iwanowicz, D. Iwanowicz , D. Smith, J. Young, J. Hedrick, S. Foster, and S. Reeser. 2007. Intersex (testicular oocytes) in smallmouth bass from the Potomac River and selected nearby drainages. *J. Aquat. Anim. Health.* 19:242-53.

Bradbury, S.P., T.C.J. Feijtel, and C.J. Van Leeuwen. 2004. Meeting the scientific needs of ecological risk assessment in a regulatory context. *Environ. Sci. Technol.* 38: 463A-470A.

Brian, J., C. Harris, M. Cholze, A. Kortenkamp, P. Booy, M. Lamoree, G. Pojana, N. Jonkers, A. Comini, and J. Sumpter. 2007. Evidence of estrogenic mixture effects on the reproductive performance of fish. *Environ. Sci. Technol.* 41, 337-344.

Caldwell, D.J., F. Mastrocco, T.H. Hutchinson, R. Lange, D. Heijerick, C. Janssen, P.D. Anderson, and J.P. Sumpter. 2008. Derivation of an aquatic predicted no-effect concentration for the synthetic hormone, 17 α-Ethynl Estradiol. *Environmental Science and Technology* 42: 7046-7054.

Daughton, C. and T. Ternes. 1999. Pharmaceutical and personal care products in the environment: agents of subtle change? *Environ. Health Persp.* 107: 907-938.

Diamond, J., J. Butcher, S. Klaine, J. Bearr, M. Bowersox, H. Latimer, T. Huang, and H. Zahner. 2006. Validation of Underlying Assumptions for Integrating Frequency, Magnitude, and Duration in Permit Conditions. 02-WSM-3, Water Environment Research Foundation, Alexandria, VA.

Diamond, J., J. Stribling, M. Bowersox, and H. Latimer. 2008. Evaluation of Effluent Toxicity as an Indicator of Aquatic Life Condition in Effluent-Dominated Streams: A Pilot Study. *Integ. Environ. Assess. Mgmt* 4: 456-470.

Diamond, J., K. Thornton, K. Munkittrick, K. Kidd, and S. Bartell. 2010a. Proceedings of Expert Workshop: Diagnostic Tools to Evaluate Impacts of Trace Organic Compounds on Aquatic Populations and Communities. CEC5R08, Water Environment Research Foundation, February 2010, Alexandria, VA.

Diamond, J., H. Latimer, K. Munkittrick, K. Kidd, S. Bartell, and K. Thornton. 2010b. Deliverable 1A: Prioritization framework for trace organic compounds (TOrCs). CEC5R08, Water Environment Research Foundation, May 2010, Alexandria, VA.

Diamond, J., J. Flippin, K. Munkittrick, K. Kidd, S. Bartell, K. Thornton, and K. Kapo. 2010c. Task 2 report: Development of Diagnostic Tools for Trace Organic Compounds and Multiple Stressors. CEC5R08, Water Environment Research Foundation, June 2010. Tetra Tech, Inc., Owings Mills, MD.

Diamond, J., K. Munkittrick, K. Kidd, S. Bartell, and K. Thornton. 2010d. Task 3 report: Testing diagnostic tools for trace organic compounds and multiple stressors: Case studies. CEC5R08, Water Environment Research Foundation, September, 2010, Alexandria, VA.

Dunlavey, E., D. Tucker, and J. Ervin. 2010. Environmental fate and transport of microconstituents. Water Environ. Technol. 22: 43-46.

ECETOC. 2007. Intelligent Testing Strategies in Ecotoxicology: Mode of Action Approach for Specifically Acting Chemicals. European Centre for Ecotoxicology and Toxicology of Chemicals. Technical Report No. 102. ISSN-0773-8072-102, Brussels, December 2007.

Furlong, E., J. Gray, D. Quanrud, S. Teske, K. Esposito, J. Marine, W. Ela, B. Stinson, D. Kolpin, and P. Phillips. 2010. Fate of estrogenic compounds during municipal sludge stabilization and dewatering. 04-HHE-6, Water Environment Research Foundation Report, Alexandria, VA.

Gasser, L., K. Fenner, and M. Scheringer. 2007. Indicators for the Exposure Assessment of Transformation Products of Organic Micropollutants. *Environ. Sci. Technol.* 41: 2445-2451.

Guo, Y., S. Krasner, S. Fitzsimmons, G. Woodside, and N. Yamachi. 2010. Source, Fate, and Transport of Endocrine Disruptors, Pharmaceuticals, and Personal Care Products in Drinking Water Sources in California. NWRI Final Project Report. National Water Research Institute, Publication Number NWRI-2010-02. Fountain Valley, CA.

Hannah, R., V. D'aco, P. Anderson, M. Buzby, D. Caldwell, V. Cunningham, J. Ericson, A. Johnson, N. Parke, J. Samuelian, and J. Sumpte. 2009. Exposure assessment of 17a-ethinylestradiol in surface waters of the United States and Europe. *Environ. Toxicol. Chem.* 28: 2725-2732

Helbling, D., J. Hollender, H-P Kohler, H. Singer, and K. Fenner. 2010. High-Throughput Identification of Microbial Transformation Products of Organic Micropollutants *Environ. Sci. Technol.* 44: 6621-6627.

Kern, S., K. Fenner, H. Singer, R. Schwarzenbach, and J. Hollender. 2009. Identification of Transformation Products of Organic Contaminants in Natural Waters by Computer-Aided Prediction and High Resolution Mass Spectrometry. *Environ. Sci. Technol.* 43: 7039-7046 .

Kidd, K.A., P.J. Blanchfield, K.H. Mills, V.P. Palace, R.E. Evans, J.E. Lazorchak, and R.W. Flick. 2007. Collapse of a fish population after exposure to a synthetic estrogen. *Proc. Nat. Acad. Sci.* 104: 8897-8901.

Klecka, G, C. Persoon, and R. Currie. 2009. Review of Contaminants of Emerging Concern and Analysis of Environmental Exposures in the Great Lakes Basin. International Joint Commission of the Great Lakes.

Kolpin, D.W., E.T. Furlong, M.T. Meyer, E.M. Thurman, S.D. Zaugg, L.B. Barber, and H.T. Buxton. 2002. Pharmaceuticals, hormones, and other organic wastewater contaminants in U.S. Streams, 1999-2000: A national reconnaissance. *Environmental Science and Technology* 36(6): 1202-1211.

Kortenkamp A. and R. Altenburger. 1999. Approaches to assessing combination effects of oestrogenic environmental pollutants. *Science Total Environment* 233:131-140.

Ort, C. M. Lawrence, J. Rieckermann, and A. Joss. 2010. Sampling for Pharmaceuticals and Personal Care Products (PPCPs) and Illicit Drugs in Wastewater Systems: Are Your Conclusions Valid? A Critical Review. *Environ. Sci. Technol.* 44: In press.

Pait, A.S. and J.O. Nelson. 2002. Endocrine Disruption in Fish: An Assessment of Recent Research and Results. NOAA Tech. Memo. NOS NCCOS CCMA 149. Silver Spring, MD: NOAA, NOS, Center for Coastal Monitoring and Assessment 55pp.

Phillips, P., S. Smith, D. Kolpin, S. Zaugg, H. Buxton, E. Furlong, K. Esposito, and B. Stinson. 2010. Pharmaceutical formulation facilities as sources of opioids and other pharmaceuticals to wastewater treatment plant effluents. *Environ. Sci. Technol.* 44:4910-4916.

Poynton, H.C. and C.D. Vulpe, 2009. Ecotoxicogenomics: Emerging Technologies for Emerging Contaminants. *J. Amer. Water Res. Assoc.* 45:83-96.

Ruby, S.M. R. Hull, and P. Anderson. 2000. Sublethal lead affects pituitary function of rainbow trout during exogenous vitellogenesis. Arch. *Environ. Contam. Toxicol.* 38:46-51.

Sandström, O. and E. Neuman. 2003. Long-term development in a Baltic fish community exposed to bleached pulp mill effluent. *Aquatic Ecology* 37:267-276.

Snyder, S., P. Westerhoff, Y. Yoon, and D. Sedlak. 2003. Pharmaceuticals, personal care products, and endocrine disruptors in water: implications for the water industry. *Environ. Eng. Sci.* 20:449-456.

Suter, G.W., II, S.B. Norton, and S.M. Cormier. 2002. A methodology for inferring the causes of observed impairments in aquatic ecosystems. *Environ. Toxicol. Chem.* 21(6):1101-1111.

Teske, S. and R. Arnold. 2008. Removal of natural and xeno-estrogens during conventional wastewater treatment. Rev. *Environ. Sci. Biotechnol.* 7:107-124.

Thomas, P. 1988. Reproductive endocrine function in female Atlantic croaker exposed to pollutants. *Mar. Environ. Res.* 24:179-183.

Thoresson G. 1996. Guidelines for coastal fish monitoring. Swedish National Board of Fisheries, Institute of Coastal Research, Kustrapport 1996:2.

IWA Publishing's authorised EU representative for General Product Safety Regulations is Diane D'Arras, 15 rue Duret, 75116 Paris, France, e-mail: safety@iwap.co.uk.

Printed and bound by CPI Group (UK) Ltd, Croydon, CR0 4YY
12/05/2026
02108893-0013